Leckie×Leckie

SECOND EDITION **2**

HIGHER
Human Biology
course notes **2nd edition**

×— Andrew Morton —×

Text © 2007 Andrew Morton
Design and layout © 2007 Leckie & Leckie
Cover photo © Stockbyte / Getty images
Cover design by Caleb Rutherford

03/091009

ISBN 978-1-84372-489-6

Published by
Leckie & Leckie Ltd, 4 Queen Street, Edinburgh EH2 1JF
Phone: 0131 220 6831 Fax: 0131 225 9987
enquiries@leckieandleckie.co.uk   www.leckieandleckie.co.uk

Special thanks to
Project One Publishing Solutions, Edinburgh (Project management and editing)
The Partnership Publishing Solutions (Design and page layout)
Bruce Ryan and Frances McDonald (illustrations)

A CIP Catalogue record for this book is available from the British Library.

Leckie & Leckie makes every effort to ensure that all paper used in our books is made from wood pulp obtained from well-managed forests, controlled sources and recycled wood or fibre.

® Leckie & Leckie is a registered trademark.
Leckie & Leckie is a division of Huveaux PLC

Acknowledgements
Leckie & Leckie has made every effort to trace all copyright holders.
If any have been inadvertently overlooked, we will be pleased to make the necessary arrangements.
We would like to thank the following for permission to reproduce their material:
SQA for permission to reproduce past examination questions (answers do not emanate from SQA).

CONTENTS

Unit 3: Behaviour, population and the environment

Tips for the exam

Answers to quick questions

Glossary

INTRODUCTION

THE SYLLABUS

This textbook is designed to cover the whole of the SQA National Qualification: Higher Human Biology. Your teacher/lecturer may give you a copy of the syllabus, but if not, you can download one from the SQA Website by following these steps:

1 Go to: www.sqa.org.uk
2 Click on the orange NQ logo on the right-hand side of the screen.
3 Select **Human Biology** from the **Subjects** list on the left-hand side of the screen.
4 Click on 'Arrangements Documents' under the heading **Subject-specific Information**.
5 Click on 'Higher Arrangements' under the heading **Downloads**.
6 Print out pages 8–43. Don't print out the whole document because it is 81 pages long

The syllabus is written for teachers, not candidates, so the language can be a bit heavy going.

If you want a simplified syllabus, buy a copy of *Questions in Higher Human Biology*, published by Leckie & Leckie, which has a simplified syllabus summary on each page. In addition, this booklet contains many questions to test your knowledge.

THE EXAMINATION

The 160-hour course is divided into three units.

To obtain a pass in Higher Human Biology candidates must, for each unit, pass a short unit test of 40 marks (pass mark 26). They must also carry out, and write up, one practical investigation.

In addition, candidates must pass a 2½ hour course examination of 130 marks, which is offered by the SQA in May of each year. Passes are graded A, B and C.

The course examination consists of three sections:

Section A	Thirty multiple-choice questions for 30 marks
Section B	Around 70 short-answer questions for 80 marks
Section C	Two extended-answer questions of 10 marks each, from a choice of four.

In addition to testing knowledge, the SQA National Examination and the three Unit tests require candidates to carry out calculations, solve problems, deal with data and comment on unfamiliar experimental situations.
Around 25% of marks are allocated to testing these skills.

Candidates should be able to:

- select information from tables, graphs, charts, keys and diagrams
- present information in a variety of forms, including tables and graphs
- calculate percentages, averages and ratios
- plan and design procedures to test hypotheses
- identify variables and controls in experimental situations
- evaluate unfamiliar experiments by commenting on controls, control of variables, limitations of equipment and possible sources of error
- draw valid conclusions from unfamiliar experimental situations
- make predictions based on evidence.

Practising for the examination

The best way to prepare for the examination is to try some past papers.
You can obtain a free copy of a specimen examination from the SQA website.
Once you've completed step 4 on p5, click on 'Higher Specimen Question Paper' and you will obtain a full paper, plus marking scheme.

It is also worth visiting the 'Understanding Standards' website (www.understandingstandards.org.uk) and clicking on 'Human Biology'.
Look for the little 'Index' box and click on 'Higher' then on 'Marking Guidelines' and you will find lots of useful information about how the paper is marked.

You can then go on to see how different questions are marked in the examination.

However, the best source of information about past papers is the Leckie & Leckie book *Official SQA Past Papers*, which is published annually. This useful book contains the most recent five national examinations in Higher Human Biology along with clear marking schemes.

HOW TO USE THIS BOOK

Note that words which are likely to be examined appear in **bold** type the first time they appear in this book. These words appear in the Glossary on pages 100-111 of this book and also in the index of the companion text *Questions in Higher Human Biology*.

At the end of each main topic there is a set of Quick Questions, which you can use as basic revision, as you complete each topic or as you revise for the exam (or both). Answers to the questions are given on pages 95-99.

UNIT 1: CELL FUNCTION AND INHERITANCE

This 40 hour Unit contains information on cell structure, cell metabolism, the activity of enzymes, chemical (tissue) respiration, cell transport, cell defence systems, immunity, DNA replication, protein synthesis, meiosis and genetics.

THE CELL

The human body is made of many millions of cells, such as nerve cells, blood cells, muscle cells, bone cells and sex cells, each type having a specific function. All cells contain a fluid **cytoplasm** surrounded by a membrane called the **plasma membrane**. In the cytoplasm are many tiny **organelles** and a vast number of different chemical compounds. The chemical reactions which take place in the organelles and in the cytoplasm are referred to as the cell's **metabolism** and many of these metabolic reactions are **catalysed** (speeded up) by enzymes. A series of such reactions is called a **metabolic pathway**.

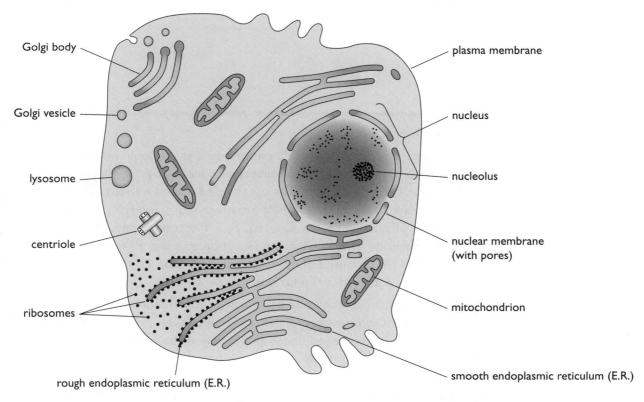

Golgi body
Golgi vesicle
lysosome
centriole
ribosomes
rough endoplasmic reticulum (E.R.)

plasma membrane
nucleus
nucleolus
nuclear membrane (with pores)
mitochondrion
smooth endoplasmic reticulum (E.R.)

A typical cell (as seen under an electron microscope)
This diagram is magnified around 2000×. The cell is invisible to the naked eye and measures around 50 micrometres. One micrometre (1 μm) is a thousandth of a millimetre or 10^{-6} m.

Each of the organelles has a specific function, as outlined in the table below:

ORGANELLE	BRIEF OUTLINE OF WHAT IT DOES
nucleus	contains DNA which acts as blueprint for the manufacture of proteins
nucleolus	involved in the synthesis of ribosomes
mitochondrion	site of production of energy-rich ATP by aerobic respiration
Golgi body	packaging of complex molecules for secretion by the cell
Golgi vesicle	small sac to enclose and transport complex molecules
rough E.R.	sheets of membrane covered in ribosomes and involved in the synthesis and transport of proteins
smooth E.R.	sheets of membrane with no ribosomes; involved in the synthesis and transport of lipids
ribosomes	found on the E.R. and free in the cytoplasm, they are the site of protein synthesis
centrioles	two cylindrical structures involved in the formation of the spindle fibres during mitosis and meiosis
lysosome	a sac derived from Golgi vesicles which contains digestive enzymes

The electron microscope

The electron microscope (EM) was invented at the same time as the television, some 70 years ago, using similar technology developed in the television cathode ray tube. The best a good optical microscope can magnify is around 1500x, but EMs are much more powerful and can magnify half a million times. Consequently, they can distinguish between points only one nanometre apart. (A nanometre is a millionth of a millimetre.) At this level of magnification, a full stop on this page would appear to be around 200 metres across.

As a consequence, many tiny structures, invisible under the optical microscope, can be viewed under an EM. However, there is one major drawback: objects viewed under an EM have to be kept in a vacuum. So, cells go through a complex preparation process which kills them. What's more, artificial changes can be brought about, called artefacts, leading to difficulties in deciding whether what you see is an accurate representation of what actually exists in a living cell.

ENZYMES

Enzymes are **proteins** manufactured by the ribosomes. Some enzymes are exported from the cell to act in, for example, the gut or the tears. Others are kept within the cell to catalyse the cell's metabolism. Without enzymes, the body's metabolism would be too slow at body temperature (37°C).

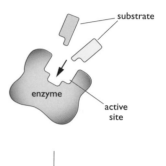

Enzymes act by combining with the **substrate** molecule(s) in a **highly specific** way so that one enzyme can normally only combine with one set of substrates. The substrates may be broken down (catabolism) or combined with others to make more complex molecules (anabolism). The substrates fit almost perfectly into an **active site** on the enzyme molecule rather in the same way as a key fits in a lock. When the reaction has taken place the fit is not quite so good and the product leaves the active site free for another reaction to take place. Some enzymes can **synthesise** or break down many hundreds of molecules per second, but their activity relies on the shape of the active site being maintained. If the temperature or pH of the surroundings vary from the **optimum** (ideal), then the enzyme molecule becomes **denatured** (altered) and no longer functions so effectively.

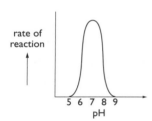

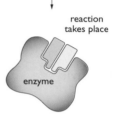

Graphs to show the effect of changes in pH, temperature, substrate concentration and enzyme concentration on the rates of enzyme reactions

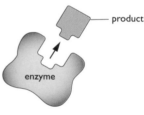

A synthesis reaction

Some compounds, which have shapes similar to enzyme substrates, compete with substrates for the active sites on the enzyme molecules. Such substances tend to slow down the activity of the enzyme and are called **competitive inhibitors**. Other types of inhibitor become attached to other parts of the enzyme molecule and alter the whole structure of the enzyme so that its active sites no longer function. Poisons such as cyanide, mercury and lead work in this fashion and are called **non-competitive inhibitors**.

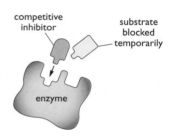

Some enzymes do not function properly unless **activators** or **coenzymes** are present. Many **vitamins** and **minerals** act as activators; for this reason they are essential in the diet. In the gut some enzymes are secreted in an inactive form because they would otherwise digest the cells which produce them. Such enzymes are activated by other enzymes in the digestive juices.

Sometimes an enzyme is absent from a cell due to a **mutation** (a fault in the genes). In such cases the metabolic pathway is blocked and very often the

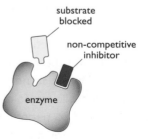

Action of inhibitors

effect is lethal. For example, the absence of the enzyme which converts the amino acid phenylalanine to the amino acid tyrosine results in a condition called **phenylketonuria** (**PKU**). The metabolic pathway is shown below.

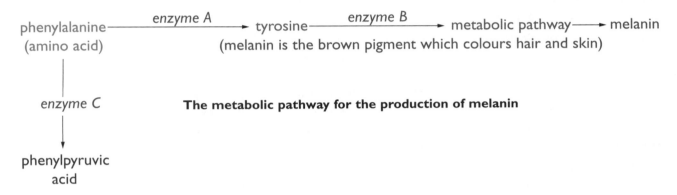

The metabolic pathway for the production of melanin

Without *enzyme A*, phenylalanine accumulates in the blood and some of it is converted into phenylpyruvic acid which is excreted in the urine. The excess of both compounds disrupts the normal development of various organs, including the brain. Fortunately, there is now a blood test which can be carried out at birth to check for this condition. Testing babies shortly after birth for genetic defects such as PKU is referred to as **post-natal screening**. Children suffering from PKU are given a low phenylalanine diet for a number of years until their brains are fully developed. Sufferers of PKU also have lighter hair and lighter skins because they can only make melanin from tyrosine and not from phenylalanine. Tyrosine is obtained in food, as are all the other amino acids.

Questions

1   What term is used to describe all the chemical reactions taking place in the body?
2   A cell is 50μm in diameter. If it is magnified 400x in a microscope, how big will it appear?
3   Where are proteins synthesised in a cell?
4   Name the organelle bounded by two membranes, the inner one of which is deeply folded.
5   What do lysosomes contain?
6   What is the function of enzymes in metabolic pathways?
7   What term is used to describe enzymes which have been damaged by high temperatures?
8   Where do chemical reactions take place on an enzyme molecule?
9   What function do many vitamins and minerals have in relation to enzyme action?
10   Name the two main classes of enzyme inhibitor.
11   Why are some gut enzymes produced in an inactive form?
12   What is melanin?
13   Why do people with PKU not produce very much melanin?
14   What term is used to describe the testing of babies shortly after birth for conditions such as PKU?

PROTEINS

Proteins are **organic compounds** composed of **hydrogen, oxygen, nitrogen** and **carbon** atoms with the addition of various other elements such as sulphur. They are complex molecules with three-dimensional structures which are unique for each protein. There are tens of thousands of different types of protein; each protein is composed of a large number of building blocks called **amino acids**, of which about twenty types are found in nature. These amino acids are polymerised (linked together) by **peptide bonds** to form chains called **peptides** or **polypeptides**. The sequence of amino acids is called the **primary structure** of the protein, and this sequence is determined by the sequence of bases in the DNA molecule. How this coding system works is described on the next two pages.

The chain of amino acids can be twisted into coils (helices), or into sheets of molecules, linked together by **hydrogen bonds**. This is the **secondary structure** of a protein. Finally, these secondary formations are twisted and bent to shape the fully-formed **tertiary structure** of the protein.

Proteins have a multitude of roles to perform in the body, so much so that it would not be far from the truth to say we are 'made of protein'.

- All enzymes are proteins.
- Many hormones are proteins. For example: ADH (antidiuretic hormone) is a peptide; growth hormone is a protein; TSH (thyroid-stimulating hormone) is a glycoprotein; prolactin is a protein.
- Muscle contraction is brought about by the sliding action of two types of protein filament: **actin** and **myosin**. (See diagram below.)
- **Haemoglobin** is a protein which transports oxygen in the blood. **Transferrin** is a protein which carries iron.
- **Antibodies** are proteins which protect the body from foreign antigens.
- **Fibrinogen** and **prothrombin** are proteins which protect the body by clotting the blood at wounds.
- All membranes have a protein component.
- The cell has a **cytoskeleton** which is composed of protein. For example, microvilli, spindle fibres and cilia all maintain their shape because of their protein 'skeletons'.
- Bones, tendons, ligaments, skin and hair are all principally composed of proteins such as collagen, elastin and keratin.

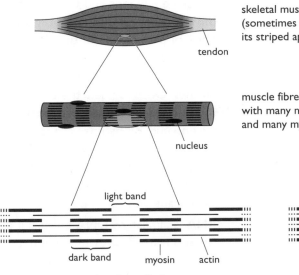

skeletal muscle
(sometimes called **striated** muscle because of its striped appearance under the microscope)

tendon

muscle fibre (cell)
with many nuclei
and many myofibrils

nucleus

light band

dark band   myosin   actin

Muscle structure   relaxed myofibril   contracted myofibril

DNA, RNA AND PROTEIN SYNTHESIS

Deoxyribonucleic acid (**DNA**) is the chemical which carries the genetic code ('blueprint') for building almost all living things, including humans. When cells are dividing, DNA can be seen under light microscopes as immensely coiled threads called **chromosomes** which are found in the nucleus. DNA is not a protein itself but it dictates which proteins are made in our cells. The DNA of our sex cells is particularly important because it carries the genetic instructions from one generation to the next. DNA is a relatively simple molecule made up of millions of **nucleotides** arranged in two spiralling rows rather like a twisted ladder.

The spiral shape of the molecule was discovered by Watson and Crick in 1952 and is described as a **double helix**. Each nucleotide is composed of a **phosphate** molecule, a **sugar** molecule and a single **base** of which there are four types. The four bases exist in pairs and make up the 'rungs' of the double helix. **Adenine** always pairs with **thymine**; **guanine** always pairs with **cytosine**. The sequence of these bases dictates the manufacture of proteins in the cell.

Proteins are made at the ribosomes, but the code for protein manufacture resides in the nucleus. So a messenger molecule is needed to carry the code from the nucleus to the ribosomes. This molecule is called **messenger ribonucleic acid (mRNA)**. The transfer of the complementary code from the DNA to the mRNA is called **transcription**. During transcription, the DNA molecule unwinds and unzips and complementary mRNA bases bind temporarily with the exposed code of part of one of the DNA strands. Start and stop codes tell the mRNA where the code to make a particular protein starts and finishes. Once formed, the mRNA strand leaves the nucleus via a nuclear pore and moves to a ribosome. Ribosomes are found free in the cytoplasm, or attached to the E.R. Free ribosomes synthesise proteins for use within the cell; attached ribosomes synthesise proteins for the cell membrane and for export.

RNA is similar to DNA but has a different base (**uracil** instead of thymine) and a different sugar (**ribose** instead of **deoxyribose**). Moreover, RNA exists as a single strand rather than a double strand.

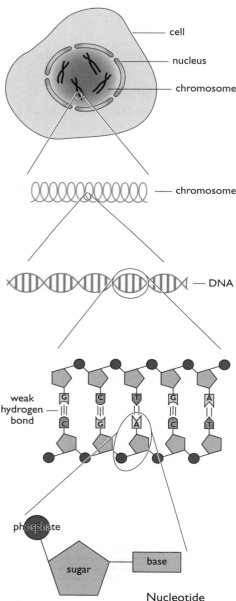

cell

nucleus

chromosome

chromosome

DNA

weak hydrogen bond

phosphate

sugar

base

Nucleotide

Chromosome structure

At the ribosomes, a second type of RNA called **transfer RNA (tRNA)** brings in amino acids for assembly into protein. tRNA exists in strands of which three particular bases, called **anti-codons**, correspond to the appropriate **triplet codons** of the mRNA.

The code is in triplets to allow for the coding of the twenty different types of **amino acid**. Since four bases can be arranged in threes in sixty-four different ways, many amino acids have more than one code, e.g. UUU and UUC both code for the amino acid phenylalanine. In addition, some codes act as 'stop' and 'start' codes. The codes for all twenty amino acids are given on *page 28* of the companion book *Questions in Higher Human Biology*.

Example of DNA transcription

DNA code	AAT-CGT-AGG
mRNA codon	UUA-GCA-UCC
tRNA anti-codons	AAU-CGU-AGG

The nine bases in the table above will code for three amino acids which will make up a small part of a peptide chain. A number of peptides linked together make a polypeptide and a number of polypeptides linked together make a protein.

The assembly of amino acids in the correct sequence, from the mRNA code, is called **translation**.

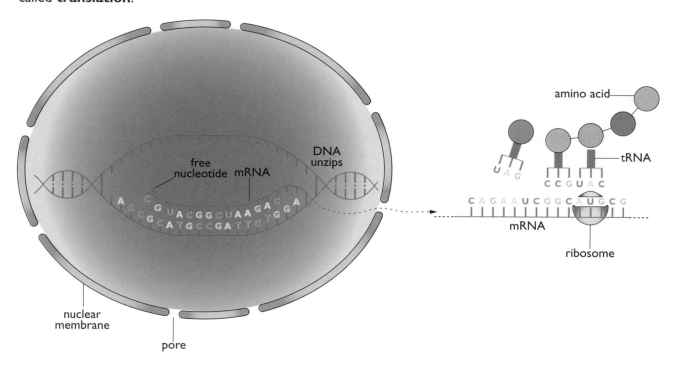

Protein synthesis

Bananas!

Tell someone you share 40% of your DNA with a banana, and they'll think you've lost your senses.

However, the claim is not as silly as it first appears.

Scientists can now decipher the genetic code of any organism on Earth, and the Human Genome Project was completed in 2003. Scientists now know the order of the 3 billion bases which make up the human genome, and DNA analysis can be used for a wide variety of purposes, for example, in medicine, crime fighting and for tracing evolutionary history.

By comparing the DNA of different organisms it is possible to estimate their evolutionary relatedness. The fact that we share over 98% of our genes with chimpanzees is not surprising, but how can we share any genes with a banana? The answer is quite simple: our cells are designed and operate in almost exactly the same way. We share a nucleus, cell membranes, ribosomes, mitochondria, ER and Golgi bodies. Moreover, processes such as DNA replication, protein synthesis and respiration are virtually identical. It takes a lot of genes to make a cell, so it's not really surprising after all. What it does tell us, though, is that we shared a common single-celled ancestor many hundreds of millions of years ago.

Quick Questions

15 What four chemical elements are present in all protein molecules?
16 What are the basic molecular building blocks from which proteins are made?
17 What bonds link amino acids?
18 What type of bond produces the secondary foldings of protein molecules?
19 Which of the following are not proteins?
    amylase   human immunoglobulin antibody   ATP   actin   DNA
20 What happens to actin and myosin proteins to produce muscle contraction?
21 Name the sugar molecule found in DNA.
22 What are the basic molecular building blocks from which DNA and RNA are made?
23 Which base pairs with adenine?
24 Which base pairs with cytosine?
25 Which base replaces thymine in RNA molecules?
26 What is a codon?
27 A single strand of DNA contains 36 bases. What is the maximum number of amino acids this could code for?
28 What is the function of mRNA?
29 What is the function of tRNA?
30 What is the mRNA complement of the DNA code ACT?
31 What name is given to a short chain of amino acids?

ENERGY TRANSFER

All living things respire, as it is this chemical process which liberates the energy in food to enable life functions such as growth, repair and movement to take place. The process is sometimes referred to as **chemical** or **tissue respiration** to distinguish it from its common usage to mean 'breathing'.

Respiration involves the **oxidation** of glucose to carbon dioxide and water, although other substances, such as fats, can also be oxidised. This transfers the chemical energy of the foodstuff to another energy-rich compound called **adenosine triphosphate (ATP)** which can be used in metabolic reactions as an immediate source of energy. If glucose solution is added to fresh muscle tissue nothing happens; if ATP is added, the actin and myosin fibres slide closer together and the muscle contracts.

ATP is constantly manufactured in all living cells from **adenosine diphosphate (ADP)** and **inorganic phosphate (P_i)**. The rate of production of ATP varies to meet the demands set by the cell. So the mass of ATP in the body remains fairly constant at all times.

The process of respiration starts in the cytoplasm and continues in the mitochondria if oxygen is present. There are many small steps in this **aerobic** (with oxygen) process to ensure the energy is

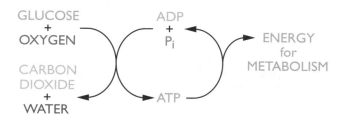

released in a gradual and controlled way. In essence, hydrogen is removed from glucose ($C_6H_{12}O_6$). This leaves an excess of carbon and oxygen atoms which are released to the atmosphere eventually as carbon dioxide. The hydrogen is then combined with oxygen from the atmosphere to make water. As water is formed, a large quantity of energy becomes available for the synthesis of ATP, so much so that up to 38 molecules of ATP can be produced from the complete oxidation of only one molecule of glucose.

If no oxygen is available, the mitochondria are not involved and the respiration process is said to be **anaerobic**. This process is much less efficient, as only 2 molecules of ATP are produced for each molecule of glucose metabolised. What is more, the end product, lactic acid is toxic, and has to be removed later by the body, using oxygen. This happens after heavy exercise and is referred to as 'repaying the oxygen debt'.

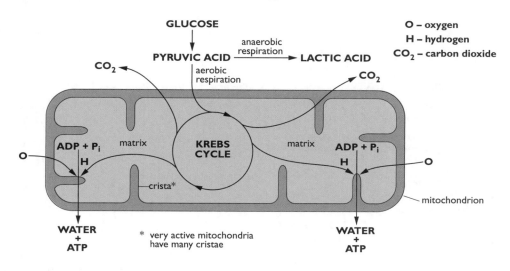

Brief summary of chemical respiration

ANAEROBIC AND AEROBIC RESPIRATION

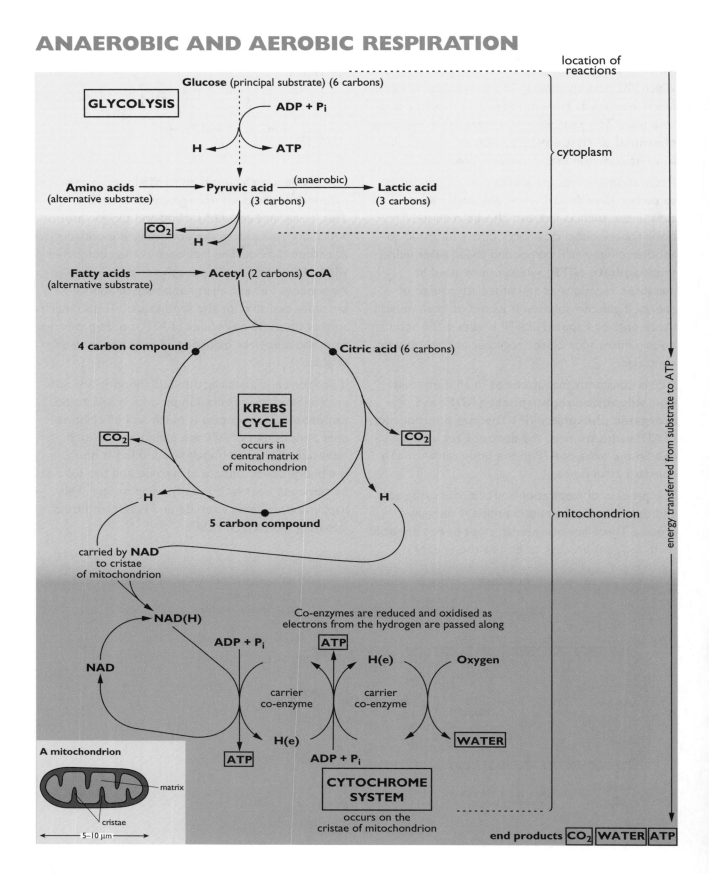

location of
reactions

Glucose (principal substrate) (6 carbons)

GLYCOLYSIS

ADP + P$_i$

H

ATP

cytoplasm

Amino acids
(alternative substrate)

Pyruvic acid
(3 carbons)

(anaerobic)

Lactic acid
(3 carbons)

CO_2

H

Fatty acids
(alternative substrate)

Acetyl (2 carbons) **CoA**

4 carbon compound

Citric acid (6 carbons)

**KREBS
CYCLE**

occurs in
central matrix
of mitochondrion

CO_2

CO_2

H

H

5 carbon compound

carried by **NAD**
to cristae
of mitochondrion

mitochondrion

NAD(H)

Co-enzymes are reduced and oxidised as
electrons from the hydrogen are passed along

ADP + P$_i$

ATP

NAD

H(e)

Oxygen

carrier
co-enzyme

carrier
co-enzyme

H(e)

WATER

ATP

ADP + P$_i$

A mitochondrion

matrix

cristae

5–10 μm

**CYTOCHROME
SYSTEM**

occurs on the
cristae of mitochondrion

end products CO_2 WATER ATP

energy transferred from substrate to ATP

SOURCES OF ENERGY

Metabolism is the sum total of all the chemical reactions taking place in the body and can be subdivided into two categories:

- anabolic reactions cause simple molecules to combine to make complex molecules and require the input of energy
- catabolic reactions cause complex molecules to break down to simple molecules with the release of energy. All chemical reactions involve the transfer of energy.

Many compounds are used as sources of energy in the cell. Proteins, fats and carbohydrates can all be oxidised to produce ATP. However, glucose, a **monosaccharide carbohydrate**, is the most common respiratory substrate. Monosaccharides are the building blocks of all carbohydrates and are often called 'simple sugars'. They contain a number of carbon atoms combined with hydrogen and oxygen atoms in the ratio of 2:1, as in water (H_2O) – hence the name, carbohydrate. Their general formula can be summarised $[CH_2O]_n$ where n is the number of carbon atoms in the sugar.

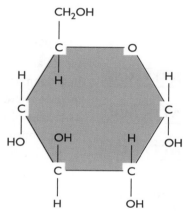

A glucose molecule $C_6H_{12}O_6$

The most important monosaccharides in the body are glucose, fructose and galactose (with 6 carbons each); deoxyribose and ribose (with 5 carbons each). Fructose and galactose are converted by the liver to glucose. Deoxyribose and ribose are components of DNA and RNA nucleotides.

If two monosaccharides are combined, a **disaccharide** is formed. The commonest disaccharide is table sugar (properly called **sucrose**) which is composed of glucose and fructose combined together. Milk sugar (**lactose**) is a disaccharide composed of glucose and galactose. Malt sugar (**maltose**) is a disaccharide composed of two molecules of glucose.

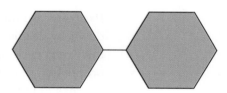

A disaccharide

When many monosaccharides are combined together, large insoluble **polysaccharides** are formed, such as starch, glycogen and cellulose. These three are all composed of many glucose molecules combined in a variety of ways. Since these molecules are insoluble, they are very useful as stores of glucose because they have no osmotic effect on their surroundings. Highly concentrated solutions of glucose would draw water out of cells and destroy them. Humans, like many other animals, store glucose as **glycogen** in the muscles and the liver. Excess glucose can also be stored as fat (a **lipid**).

A polysaccharide

Lipids liberate twice as much energy as carbohydrates (per unit mass) but their breakdown takes longer. So in a marathon race – when glycogen stores are exhausted and only fats left as a source of energy – the runner 'hits the wall' and slows down to a crawl, totally exhausted, but not in fact out of energy. There is still plenty of fat available – it is just a slow process transporting and metabolising fat to generate ATP. Interestingly, the heart, unlike other muscles of the body, has a preference for fat as a source of energy. If the heart relied solely on glucose then the exhausted runner would have died! Experienced marathon runners load up with carbohydrate before the event by altering their diet, then pace the race so that there is just sufficient glucose and glycogen to maintain top speed to the finishing line.

Lipids are similar to carbohydrates, in that they are composed of carbon, hydrogen and oxygen, but the atoms are present in different proportions. They include waxes, oils, fats and steroids. Like polysaccharides, they are very useful as stores of energy because they are insoluble in water. But they have many other important functions:

- Nerve fibres are insulated by a fatty myelin sheath (see *page 61*).
- Our feet and hands are protected by pads of fat.
- Phospholipids are essential components of all membranes (see *page 20*).
- Vitamins A and D, which are insoluble in water, are transported by fats.
- Sebum, a wax, waterproofs our skin and protects it from bacteria.
- Steroids, such as oestrogen and testosterone, are important hormones (see *pages 37-39*).
- Cholesterol, another steroid, is also essential for membrane construction.
- A layer of fat under the skin reduces heat loss from the body.

Proteins are not stored in the body. So, when excess proteins are broken down, by deamination (see *page 55*) in the liver, some of the products can enter the Krebs cycle and be broken down to release energy. This provides a small proportion of our daily energy requirement, but is not important. However, in times of extreme starvation when carbohydrate and fat reserves have been almost used up, the body will start to break down its own structural proteins, such as actin and myosin, to obtain energy. Eventually, when the heart muscle is broken down, death results.

GLYCEROL — FATTY ACID

GLYCEROL — FATTY ACID

GLYCEROL — FATTY ACID

A simple fat molecular (triglyceride)

COMPOUND	HOW IT IS USED AS A SOURCE OF ENERGY
ATP	Immediate source of energy to drive all metabolic reactions
Glucose	Principal substrate for generation of ATP during respiration
Glycogen	Principal carbohydrate store in liver and muscles. Can be converted to glucose
Fat	Principal long term energy store, but rather slow to metabolise
Protein	Needed for body building, so only used as a last resort as a source of energy

Questions

32 What is the function of chemical respiration?

33 What energy-rich compound is produced as a consequence of chemical respiration?

34 What is the most common respiratory substrate?

35 Name an alternative respiratory substrate.

36 How many molecules of ATP can be synthesised during the aerobic breakdown of one molecule of glucose?

37 Where does glycolysis take place in the cell?

38 Name the three-carbon end-product of glycolysis.

39 Name the second stage of respiration which occurs in the matrix of the mitochondrion.

40 Name the gas released during the second stage of respiration.

41 What is the function of NAD?

42 Where precisely are the cytochrome carrier co-enzymes found in a cell?

43 Name the final hydrogen acceptor in the respiratory system.

44 Name a useful end-product of aerobic respiration, apart from ATP.

45 What compound is removed from your body when you 'repay your oxygen debt'?

46 Which of the following carbohydrates is a monosaccharide?

    maltose   **glucose**   **sucrose**

47 Which of the following carbohydrates is a disaccharide?

    maltose   **glucose**   **fructose**

48 Why are insoluble polysaccharides useful for storing energy in the body?

49 Where is glycogen stored in the body?

50 What is the collective term for substances such as fats, oils and steroids?

51 How do these substances compare with carbohydrates in terms of energy content?

52 What two types of molecule are found in triglycerides?

53 Which hormones are referred to as steroids?

54 Which substance is used last as a source of energy, when someone is starving?

CELL MEMBRANES

Cells are full of membranes and are surrounded by a membrane called the **plasma membrane**. Many organelles (such as mitochondria, endoplasmic reticulum and the nucleus) are bounded by, or composed of, membranes.

Membranes are exceedingly thin and are made of **phospholipids** and **proteins**. The lipids form two layers of molecules which are mobile. The proteins are found scattered as a mosaic in and on the lipid layers and they too can move around the membrane.

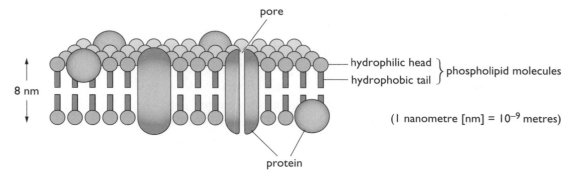

For this reason, the membrane is described as a **fluid-mosaic**. Membranes keep their shape because of the hydrophobic and hydrophilic nature of their lipid molecules.

One end of the molecule is repelled by water (hydrophobic) and the other is attracted to water (hydrophilic). As a consequence of their fluidity, membranes can often recover from minor physical damage. Lipids also allow small molecules, such as water molecules, to pass through the membrane unaided. The surface area of cells or organelles can be increased by the folding of membranes. Examples are the microvilli (see *page 54*) of cells lining the gut and the cristae of the mitochondria.

Because they are composed of proteins and lipids, membranes are easily damaged by heat, acids and by fat solvents such as alcohol. This can be demonstrated in the laboratory using cells from beetroot or red cabbage which have a coloured sap. When the membrane of these cells is damaged the coloured sap leaks out into the surrounding water.

Membrane proteins have many functions:

● There are enzymes present in membranes for many chemical reactions which take place on the surface of the membranes, e.g. stages in protein synthesis and respiration.

● There are receptor sites for hormones which then influence the activity of the cell.

● There are proteins which act as a skeleton for the membrane to give it shape and to allow it to move. For example, the membrane moves during endocytosis and exocytosis and when cilia waft to and fro.

- There are proteins which actively transport materials across the membrane using ATP as a source of energy – this is called **active transport** (see below).
- There are proteins which form pores in the membrane through which substances can pass.
- There are proteins which act as markers for self-recognition (see *page 31*).

MEMBRANE TRANSPORT

Because cells are full of hundreds of different enzymes and millions of molecules which react every second, it is important that the internal environment of the cell is kept as constant as possible. In this respect, the plasma membrane plays a very important function by regulating what enters and leaves the cell. It allows some substances to pass through unaided, while others are transported actively and yet others are entirely prevented from passing through. For this reason membranes are said to be **selectively permeable**.

Atoms and molecules of gases and liquids move about at random. As a result they tend to spread themselves from areas of high concentration to areas of low concentration. This process is called **diffusion**. **Osmosis** is the term used to describe the diffusion of water across a membrane from a weak solution (high proportion of water molecules) to a strong solution (low proportion of water molecules). Cells placed in fresh water tend to burst because water diffuses into them. Cells placed in a strong salt or sugar solution shrink as water diffuses out of them. So it is important that our bodies maintain a constant osmotic environment in the tissue fluids.

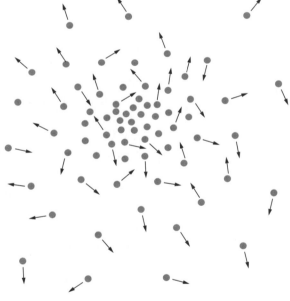

Diffusion

Some membrane proteins allow molecules to pass through under their own 'power' by diffusion. But, if a cell needs to move a substance against the **concentration gradient**, it must supply energy in the form of ATP and transport the substance **actively**. There are proteins in membranes which act as pumps to force substances, such as mineral ions, from one side of the membrane to the other against the concentration gradient. Because **active transport** requires energy supplied by the cell, it is affected by the rate of respiration taking place in the cell. So factors such as the temperature, availability of oxygen and glucose, and the presence of enzyme inhibitors all affect rates of active transport.

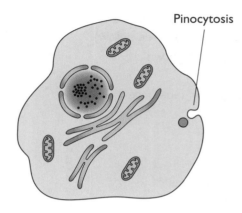

Pinocytosis

Pinocytosis

Large insoluble molecules (e.g. proteins) and larger items (e.g. whole bacteria) cannot pass through the membrane as such, but they can be transported from one side to the other in packages of membrane. The membrane flows round the item to form a tiny vacuole (vesicle) and the item is moved into (**endocytosis**) or out of (**exocytosis**) the cell. If large solid molecules are engulfed by the cell on a grand scale, the process is called **phagocytosis** (*see below*). If fluid is engulfed by the membrane, the process is called **pinocytosis**.

IMMUNITY

Throughout our lives, our bodies are constantly bombarded by foreign micro-organisms and the poisons they produce. To protect ourselves, a wide variety of defence systems have evolved. In the first instance, we keep organisms out of our bodies by a variety of **innate** (inborn) first line defence mechanisms. For example, the epithelial cells of the skin, the mucus membranes of the lungs and the gut, the ciliated cells of the respiratory tract, stomach acid, vaginal acid, skin sebum and the enzyme lysozyme of tears, all prevent, or reduce, the chances of invasion by foreign organisms. If germs break through these first line defences, the body fluids (blood plasma, tissue fluid and lymph) contain many cells of the **immune system** which can neutralise invading microbes and their toxins in a variety of ways.

A foreign substance (usually a protein or polysaccharide) which triggers a reaction from the immune system is given the general name **antigen**. Any protein which the body produces in response to an antigen invasion is called an **antibody**. But not all foreign substances provoke an attack. If foreign molecules are fairly simple, the immune system does not recognise them as being foreign and leaves them alone. For example, plastic molecules, with many simple identical repeating units, can be used to manufacture artificial implants, such as hip joints and heart valves, because they do not stimulate an immune response.

The cells of the immune system are not organised as distinct organs like other body systems but, rather, consist of billions of cells throughout the body which are able to deal with antigens in one way or another. There are so many that if they were grouped together to form a single organ, they would occupy a volume similar to that of the brain. Some cells, **macrophages**, deal with invasion in a generalised non-specific way. Macrophages are phagocytic cells found throughout the body, but

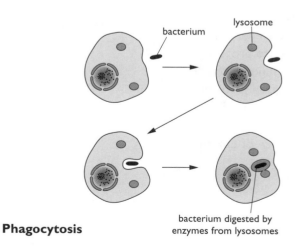

bacterium

lysosome

Phagocytosis

bacterium digested by enzymes from lysosomes

particularly in the lymph system. They engulf any foreign particles they find. Others are highly specific and each one can deal only with one type of antigen. These are the **lymphocytes**.

There are two basic types of lymphocyte, the **T-lymphocyte** and **B-lymphocyte**. Both types are white blood cells and look identical under the microscope but they mature in different regions of the body and act in quite different ways. The T-lymphocytes are so called because they mature in the thymus gland. There are a number of different types of T-lymphocytes with a variety of functions but, in essence, they attack and destroy body cells which have been infected by viruses or bacteria. These infected body cells can be identified because of markers (a carbohydrate + protein) on the plasma membrane which have been altered by the presence of the microbe. Every person has their own unique 'signature' (set of markers) on the surface of all their cells, except for identical twins (monozygotic twins – see *page 41*) who have identical markers.

During development of the fetus, the immune system learns to recognise these **'self'** markers and thereafter recognises foreign antigens as being **'non-self'**. A special binding site on the T-lymphocyte binds with the 'non-self' cell and destroys it, so destroying the microbe within. This is called the **cell-mediated response**.

The T-lymphocytes will also remove cancer cells and grafted tissue which they recognise as being foreign. For this reason, grafting of tissues or organs from one person to another is fraught with difficulty. Surgeons take care to match the antigens of each person as well as they can before transplant operations. Drugs are also used to suppress the immune system of the patient but, even then, donated organs are often rejected.

The B-lymphocytes mature in the bone marrow and protect the body by producing specific antibodies for specific antigens. The antibodies act at a distance from the parent B-lymphocyte and this is called the **humoral response**. The antibodies bind to antigens, making them temporarily inactive, and often act as markers for recognition by T-lymphocytes and phagocytic cells such as macrophages.

Very early in life we acquire millions of different T-lymphocyte and B-lymphocyte cells, ready-made, to deal with any specific antigen which might invade our bodies. Consequently, many T-lymphocytes and B-lymphocytes, which are never needed to fight infection, exist in the body. When an antigen invades, the appropriate T-lymphocytes and B-lymphocytes are already in place, but there is a delay until sufficient numbers of the same type are generated by mitotic division to deal with the invasion. However, once these cells have experienced a specific antigen, special memory cells linger in the

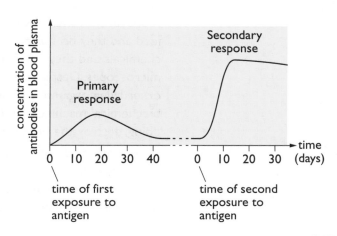

body fluids and enable the immune system to produce more antibodies more quickly if the same antigen is experienced again. In other words, we become immune to the disease.

The immune system, like any other body system, can sometimes malfunction and fail to recognise body cells as 'self'. In such cases an **autoimmune response** occurs and the immune system attacks its own body cells as if they were foreign. There are a number of diseases which are thought to arise from autoimmunity, including forms of rheumatoid arthritis, diabetes and myxoedema.

Sometimes we suffer an **allergic reaction** to relatively innocent substances such as pollen grains, animal hair, feathers or food additives. Such substances are recognised as being foreign by the immune system which overreacts in an attempt to deal with the 'antigen'. About 30% of the UK population has an allergy of one kind or another, such as hay fever, eczema and food allergies. Children often grow out of allergic reactions, but an increasing number of children and adults suffer from allergies because they are bombarded by a growing cocktail of toxic chemicals in their daily lives.

The body can become 'immune' to antigens in a variety of ways, some of which occur naturally and some of which are artificially induced. The table summarises the various forms of immunity and how they are acquired.

Types of acquired immunity

IMMUNITY	ACTIVE	PASSIVE
Natural	antibodies made by body after an infection	antibodies passed across placenta or given in breast milk
Artificial	antibodies made by body after a vaccination	antibodies given by injection/inoculation
advantages and disadvantages	slow acting but long-term protection	quick acting but only short-term protection

VIRUSES

Viruses are the smallest known living things (10–100 nm). In fact some would question whether they can be classified as living organisms. They cannot reproduce on their own, they do not grow, they do not respire, they do not feed and they do not produce waste. They can be crystallised like simple chemicals and they are invisible except under the most powerful electron microscopes. Despite their size, however, they can have profound effects on other living organisms because they are always parasitic. They make use of the biochemical machinery of host cells to make copies of themselves and, in so doing, destroy the host cells.

Viruses are made of a protein coat containing a DNA or RNA core. When they combine with host cells, they inject their DNA or RNA into the cell. The host cell then comes under the command of the new genetic instructions and makes new viruses using nucleotides, amino acids, ATP and enzymes supplied by the host cell. The viruses, once formed, burst out of the host cell, killing it in the process. For this reason we suffer symptoms of illness and may even die from a viral infection.

Many human diseases are caused by viruses, and such diseases are difficult to treat because they do not respond to the use of antibiotics. We have to leave it to our own immune system to deal with the viral attack. Human diseases such as polio, AIDS, smallpox, rabies, measles, mumps, the common cold and flu are all caused by viruses.

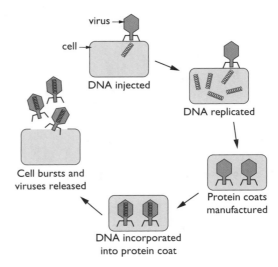

Viral replication

Quick Questions

55   What are the two principal components of membranes?
56   What term is used to describe the mobile nature of membranes?
57   Describe two functions of membrane proteins.
58   What special term is used to describe the diffusion of water molecules through a membrane?
59   What happens to human cells if they are placed in fresh water?
60   In what two ways is active transport different from diffusion?
61   Describe one feature which pinocytosis and phagocytosis have in common.
62   What is an antigen?
63   Name the white blood cells which engulf and digest bacteria.
64   Name the organelles which contain powerful digestive enzymes used to digest the bacteria.
65   What cells produce antibodies?
66   What name is given to the antibody production response?
67   What cells are involved in the cell-mediated response?
68   In what two ways is the secondary response to infection different from the primary response?
69   What term is used to describe the situation where the body sets up an antibody response against its own cells?
70   What is the difference between active and passive immunity?
71   Give an example of gaining passive immunity in a natural way.
72   Give an example of gaining active immunity in an artificial way.
73   Which two of the following compounds are found in viruses?
      carbohydrates   nucleic acids   amino acids   fatty acids   ATP

INHERITANCE

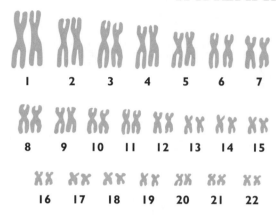

sex chromosomes
X and Y

23

A karyogram of a male

We are like our parents because we inherit their **genes**. Genes are pieces of DNA which code for every one of our characteristics. Each of our parents provides us with a complete set of genes via their respective sperm and egg. So all of our cells contain two sets of instructions for every characteristic of our body – one set from each of our parents. These pairs of alternative genes are called **alleles**. Our genes make up the **23 pairs of chromosomes** found in every body cell (**somatic** cell) except in sperm and eggs (**gametes**) and mature red blood cells. Gametes have only one set of chromosomes and red blood cells have none. The number, size and shape of these chromosomes is what we call our **karyotype**, and a diagram to show these chromosomes arranged in pairs is called a **karyogram**.

To construct a karyogram, chromosomes can be paired using the following features:

- length
- position of centromere
- banding patterns.

Before any cell divides, it must duplicate its DNA and ensure that each daughter cell contains a copy of the code. The process of copying the DNA is called **replication**, and the process of separating the copies of the DNA (chromatids) and delivering them to each new cell is called **mitosis**.

The process of replication requires a supply of:

- free nucleotides
- ATP
- the appropriate enzymes.

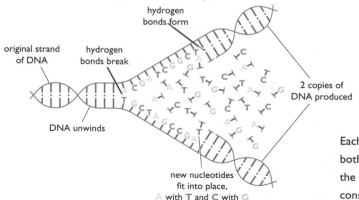

Each new double helix is composed of both an old and new strand of DNA, so the process is sometimes called semi-conservative replication.

Replication of DNA   Time

MITOSIS

At the start of mitosis, chromosomes appear as pairs of identical copies, formed from DNA replication. These chromosome replicas are called **chromatids** and they are held together at a **centromere**. As the chromosomes appear, the nuclear membrane disappears. The chromosomes migrate to the equator of the cell where the chromatids separate and are pulled apart to the poles of the cell by spindle fibres. The nuclear membrane reforms and cell division follows. Each new cell has a complete set of copies of the original chromosomes.

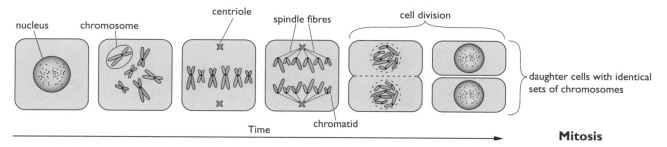

This means that all the cells of the body contain identical sets of genes. However, each somatic cell only uses those genes which are of relevance to it. For example, only red blood cells use the gene for the manufacture of haemoglobin. How this is achieved is not yet clearly understood.

MEIOSIS

All somatic cells contain **46** chromosomes. This is called the **diploid** number. However, gametes only contain **23** chromosomes, the **haploid** number, so that when a sperm fertilises an egg the full diploid number of chromosomes is regained. When gametes are formed, the process of nuclear division which reduces the diploid number to the haploid number is called **meiosis**. Also, during meiosis, genes are shuffled such that each haploid gamete which is formed has a new combination of genes. So, when the gametes fuse, the **zygote** (fertilised egg) which is formed has two unique sets of genes, one set from each parent. These matching sets of genes are on pairs of chromosomes called **homologous** chromosomes. The genes on each of these homologous pairs are found at the same points, called **gene loci**, on the chromosomes.

The shuffling, or recombination of genes during meiosis, is of enormous significance: it is the main reason why no two humans in a population of over 6 billion are identical, except for identical twins. Why is it so important that children are genetically different from one another and their parents? The reason is simple: genetic variability allows all life on Earth to survive changing conditions, and therefore to evolve more quickly. Organisms which can't adapt to changing conditions become extinct, as is only too evident in today's rapidly changing world. When the great plague swept Europe in the fourteenth century, many died, but some survived because they were genetically different. Today, AIDS is a lethal disease, but some survive the infection because they have a combination of genes which is different from everyone else.

The shuffling of the genes during meiosis results from two different events:

- the **independent assortment** of pairs of chromosomes at the equator of the cell
- the **crossing over** of pieces of chromosome at points called **chiasmata** (singular – 'chiasma').

This is shown in the diagrams below. (For clarity only four chromosomes are shown.)

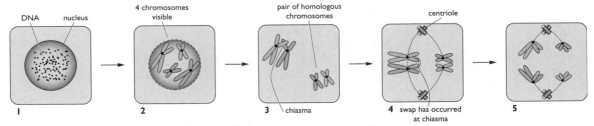

Meiosis (note: chromosomes originating from sperm are shown as shaded; chromosomes from egg are shown as unshaded)

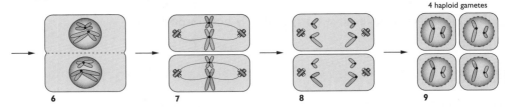

1 The chromosomes are normally invisible in the nucleus, as the DNA is not densely coiled. Before meiosis can take place, the DNA must replicate so that there are two copies of the genes to share out to the new cells which are about to be formed.

2 The chromosomes start to shorten and thicken and eventually become visible as tiny dark threads. They can be seen as pairs of **chromatids** joined at a point called the centromere. These chromatids are the identical copies of DNA formed during earlier replication.

3 The chromosomes find their homologous partners and the chromatids become intertwined at chiasmata. At these points the chromatids often break and rejoin with their opposite number and genes **cross over** from one chromatid to another. This does not happen in mitosis.

4 The homologous pairs of chromosomes migrate to the equator of the cell and line up in pairs independently of one another. Spindle fibres form at the poles and become attached to the chromosomes at their centromeres.

5 The homologous chromosomes are pulled apart towards opposite ends of the cell by the spindle fibres.

6 Once the chromosomes have reached the poles of the cell, the cell divides in two. Now each new cell has only two chromosomes and not four.

7 The two cells go on to divide again to form four sperm cells or one egg cell. (The three other potential egg cells degenerate during the process.) This second meiotic division is very similar to a mitotic division. Chromosomes line up along the equator individually.

8 The spindle fibres form and become attached to the chromatids which they pull apart.

9 Four cells are formed, each with two chromatids which can now be called chromosomes again. Some chromatids will be different from others because of the crossing over of genes at chiasmata. Moreover, the combination of chromatids in each cell will be different because of the random way in which they lined up along the equator of the cell at the first and second divisions of meiosis.

Summary of differences between mitosis and meiosis.

	MITOSIS	**MEIOSIS**
Type of cell produced	somatic	gamete
Number of cells produced	2	4
Chromosome number of cells	46 (diploid)	23 (haploid)
Genetic variability	none	considerable

Quick Questions

74 What are 'somatic' cells?

75 What term is used to describe a picture of the chromosomes of the body set out in matching pairs?

76 Describe two features used to pair chromosomes.

77 What term is used to describe these pairs of matching chromosomes?

78 Place the following events into the correct order:

> **cell division**
> **DNA replication**
> **mitosis.**

79 Name three basic substances needed for DNA replication.

80 Why is the process of replication sometimes described as semi-conservative?

81 What are the chromosome replicas called when still held together by the centromere?

82 During mitosis and meiosis what part of the cell manufactures spindle fibres?

83 Describe two important ways in which meiosis is different from mitosis.

84 At what stage in meiosis are the chromatids separated?

85 What are the diploid and haploid numbers for human cells?

86 Why is it important that gametes are haploid?

87 What term is used to describe the points at which crossovers occur during meiosis?

88 Why is it important that gametes are all genetically different?

MONOHYBRID INHERITANCE

Genuine examples of human characteristics controlled by a single pair of alleles (monohybrid inheritance) are uncommon. Some of the examples which follow are simplified for ease of understanding.

Every person has two alleles for every characteristic, one inherited from each parent. If just one of these alleles affects the growth and development of the individual on its own, it is said to be **dominant**. If an allele has no effect on an individual on its own, it is said to be **recessive**. Recessive alleles only have an effect on the development of an individual when a person inherits two recessive alleles, one from each parent. For example, **Huntington's chorea** is a particularly unpleasant condition which appears in middle age and results in early death. The allele which causes the condition is dominant. So a person with only one of these alleles will suffer from the condition and has a 50% chance of passing the allele on to his or her children. The condition **cystic fibrosis** is caused by a recessive allele. So people with one of these recessive alleles are perfectly healthy. However, if a man and woman both carry one of these alleles, there is a 25% chance that any child they have will be affected by cystic fibrosis. See the table below, which is sometimes called a Punnett square.

A^N = normal dominant allele
A^n = recessive cystic fibrosis allele

sperm / eggs	A^N	A^n
A^N	$A^N A^N$ NORMAL	$A^N A^n$ NORMAL but CARRIER
A^n	$A^N A^n$ NORMAL but CARRIER	$A^n A^n$ CYSTIC FIBROSIS

M = allele for melanin (skin pigment)
m = allele for no melanin (albino)

☐ = male   ◯ = female

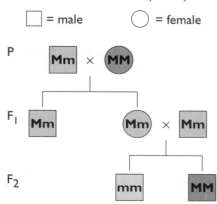

A number of technical terms are used to simplify genetic language. For example, if an individual has two identical alleles for any particular characteristic (trait) then they are said to be **homozygous**. Individuals with two different alleles are said to be **heterozygous**. The appearance of an individual with respect to a particular set of alleles is referred to as the **phenotype**. The genetic make-up of an individual is referred to as the **genotype**. Genotypes are written using letters of the alphabet. Capital letters indicate dominant alleles and small letters indicate recessive alleles. The parents are referred to as the **P** generation and thereafter the offspring are called the F_1 and F_2 generations, etc.

A family tree showing the inheritance of **albinism** in a fictitious family is shown opposite. Albinos lack the ability to make the pigment melanin which colours the skin, hair and iris. A man, who is a carrier of the recessive allele, marries a woman without the allele. They have two children, both of whom inherit the allele from their father. The chance of both children inheriting the allele is 25% ($\frac{1}{2} \times \frac{1}{2} = \frac{1}{4}$). The son does not marry but the daughter marries a carrier of the allele and they have one albino child and one unaffected child.

The chance of two carriers having an albino child is 25%. The chance of two carriers having a child who does not carry the recessive allele is also 25%.

Sometimes recessive alleles are not totally hidden by dominant alleles and the alleles are then referred to as **incompletely dominant**. In other cases, both alleles have an equal effect on the phenotype of an individual and are described as **co-dominant**. However, the distinction between incompletely dominant alleles and co-dominant alleles is not easy to quantify. Some examples are given below:

Sickle-cell anaemia is a recessive condition in which blood haemoglobin is malformed. Anyone with two 'sickle-cell' alleles suffers from severe sickle-cell anaemia. But anyone with only one 'sickle-cell' allele and one 'normal' allele suffers from a mild form of the disease. This is because only some of the haemoglobin is affected. Interestingly, those with mild sickle-cell anaemia are more resistant to malaria. The disorder is much more common in areas of the world where malaria is endemic (always present) because someone with one 'sickle-cell' allele is more likely to survive a malarial attack than someone without the allele.

Red blood cells have a number of antigenic 'self' protein markers on their surfaces which are important in situations where blood has to be transfused from one person to another. If the protein markers of a donor's blood don't match those of the recipient, the recipient will recognise the donated blood as foreign and attack it with antibodies. Coagulation can then occur and the recipient can die. One such blood group system is the ABO system. The genetics of this system are slightly more complicated because three alleles are involved. They are described as **multiple alleles**. An individual can, of course, have only two of these alleles, one from each parent. Alleles A and B are co-dominant to allele O, which is recessive. The table summarises possible genotypes and phenotypes.

BLOOD GROUPS (PHENOTYPES)	ALLELES (GENOTYPES)
A	AO or AA
B	BO or BB
AB	AB only
O	OO only
(The alleles are sometimes written I^A, I^B, I^O)	

People with blood group A have anti-B antibodies in their **plasma** (blood fluid), so they cannot receive blood from those of blood group B or AB. Likewise, people of blood group B cannot receive blood from those with blood group A or AB. People with blood group O are universal donors because they have no antigenic markers on their red blood cells; those of blood group AB are universal recipients (receivers of blood) because they have no antibodies against red blood cells in their plasma.

Another blood group system is the **Rhesus system**. Individuals who are Rhesus-positive have an antigenic marker on their cells; those who are Rhesus-negative do not. The allele for the antigenic marker is dominant (see *page 30*).

SEX LINKAGE

In humans, the sex of a person is controlled by genes found on the sex chromosomes. There are two sex chromosomes, a longer one called the **X-chromosome** and a shorter one called the **Y-chromosome**. Women have two X-chromosomes; men have an X- and a Y-chromosome. The other 44 chromosomes are referred to as **autosomes** (*see karyogram on page 26*).

egg / sperm	X	X
X	XX	XX
Y	XY	XY

XX = 50% of offspring are girls

XY = 50% of offspring are boys

During meiosis the sex chromosomes separate so that eggs can only have an X-chromosome and sperm can have a Y-chromosome or an X-chromosome. It is the sperm therefore which dictate what sex a child will be.

Not all the genes on the sex chromosomes are involved in the development of sexuality. Many have nothing to do with sex at all. These genes are said to be **sex-linked** and their inheritance does not appear to follow normal monohybrid patterns because of the lack of such genes on the short Y-chromosome. Males cannot therefore be heterozygous for such sex-linked genes. Instead, they always express these genes in their phenotype regardless of whether the genes are dominant or recessive.

egg / sperm	X^c	X^c
X^C	$X^C X^c$ carrier	$X^C X^c$ carrier
no allele Y	$X^c Y$ colour blind	$X^c Y$ colour blind

c = allele for colour blindness

C = allele for normal vision

Two examples are the genes for colour vision, and a factor involved in the clotting of blood. Their mutant recessive alleles give **colour blindness** and **haemophilia** respectively. So a colour-blind woman, having children to a normal man, will have boys who are colour-blind and girls who, although carriers of the gene, have normal colour vision, as shown in the Punnett square.

POLYGENIC INHERITANCE

Most characteristics of organisms, including humans, are controlled, not by one pair of alleles but by many pairs of alleles. This is referred to as **polygenic inheritance**. For example, if height in humans were controlled by a single pair of alleles, it would result in humans being either short or tall, rather like Mendel's pea plants. The feature would show **discontinuous** variation.

In fact, height is controlled by the combined effect of many alleles at different loci on the chromosomes. Height is also influenced by the environment, particularly by nutrition. So there exists a wide range of possible heights, and human height shows **continuous** variation. When heights or weights of a large population of humans of the same age are plotted on a graph they result in a line which shows a **normal distribution**.

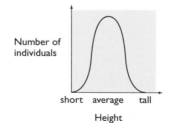

MUTATIONS

Meiosis, followed by random fertilisation, results in an almost infinite variety of gene combinations in any species of plant or animal which reproduces sexually. However, meiosis can only recombine genes which already exist – it cannot create new genes. New genes arise as a result of **mutations**.

Mutations can, and do, occur in any nucleated cells. However, only those mutations occurring in our gametes are of significance, because they are inherited. Mutations are random, undirected, spontaneous changes in DNA molecules, and can be as simple as a change in one base, or as significant as a change in large parts, or indeed numbers, of entire chromosomes in a cell.

Although mutations are the essential raw material for evolution, a vast majority are harmful. Only rarely does a mutation arise which gives an organism an improved chance of survival in a constantly changing environment. For example, the mutation in humans giving rise to haemophilia is harmful, but the mutation giving rise to sickle-cell anaemia gives heterozygotes some protection from the disease malaria.

There are many different ways in which mutations can occur. Individual bases can be added, substituted or deleted from a strand of DNA. **Substitutions** often have little effect because they change only one triplet code and might affect only one amino acid in a protein chain. However, **deletions** and **insertions** are much more serious because they alter the subsequent code for every amino acid in the chain. (See table.)

MUTATIONS	
AUG/CCC/GGC/AAU	normal mRNA
AUG/*CUC*/GGC/AAU	substitution (U for C)
AUG/*CCG/GCA/AU*...	deletion (take out C)
AUG/*CCA/CGG/CAA/U*	insertion (add A)
(affected codes in italics)	

Chromosome mutations can result in pieces of chromosome being removed (**deletion**), rotated and replaced (**inversion**), shifted from one place to another (**translocation**), or copied twice (**duplication**).

Occasionally, entire extra chromosomes arise in gametes through a failure of the spindle fibres to pull one or more of the chromosomes apart at meiosis. This meiotic failure is called **non-disjunction**. Most of these mutations are lethal but a few are not. For example, an extra autosome number 21 causes **Down's syndrome**. An extra X-chromosome in males (XXY) causes sterility, an increase in height and the development of some female characteristics such as breasts. This is called **Kleinfelter's syndrome**.

Anything which causes a mutation is called a **mutagen**. The shorter wavelengths of radiation are particularly dangerous. Alpha (α) rays, beta (β) rays, gamma (γ) rays and X-rays, which are emitted from nuclear reactions and atomic explosions, increase mutation rates, and can be lethal if of sufficient intensity. The mutation rate in humans has increased in the last sixty years due to the testing of atomic bombs and the operation of nuclear power stations. Pregnant women are rarely X-rayed, to avoid causing mutations in the developing embryonic tissues of the fetus. Moreover, people working with

X-ray machines have to be protected from the dangerous radiation. UV light and many toxic chemicals are also mutagenic. Cigarette smoke causes cancer of the lungs and sunbathing increases the risk of skin cancer. Cancer results from mutations arising in somatic cells, where these mutated cells divide and multiply in an uncontrolled way. If malignant, these cells spread throughout the body and cause new cancerous growth in other tissues.

GENETIC SCREENING AND COUNSELLING

The word 'disease' is used to describe many human conditions, but it is perhaps best reserved for those conditions which have been caused by **pathogens** (organisms which cause disease), such as viruses and bacteria, rather than those caused by faulty genes. Diseases caused by genes are perhaps better described as **genetic disorders**. Genetic disorders have a permanence about them which makes them difficult to treat; and, if there is a risk we have inherited a defective gene or genes, then we may wish to know what that risk is.

There are a number of ways of determining risk and much depends on the number of genes which affect the condition and on the condition itself. At its simplest, if only one gene is involved, as in haemophilia or cystic fibrosis, then the risks are easy to evaluate. But, if many genes are involved, the calculations become much less precise. People who advise couples on such risks are called **genetic counsellors**. They rely on family histories, and the study of cells, chromosomes and DNA, to make their predictions. For example, if a man has a father who is a haemophiliac (haemophilia is a sex-linked recessive condition), then a counsellor can tell him that he could not have inherited the gene from his father. On the other hand, a woman whose father has haemophilia will be a carrier and may pass the allele to her sons or daughters.

Duchenne **muscular dystrophy** is inherited in the same way as haemophilia except that, as sufferers die young, the condition is only passed down by unaffected carrier females. Screening tests to detect this gene before birth can be unreliable, as a third of all cases result from new mutations and are therefore born into families with no previous history of the disorder. The physical symptoms do not appear until after the age of 2 or 3. However, a family history of muscular dystrophy enables counsellors to make predictions of risk. A carrier woman has a 50% chance of having a carrier daughter or an affected son, and the daughter has a 25% chance of having an affected son.

Predictions involving polygenic disorders are much more difficult; in these cases the risks are evaluated by looking at family histories. At its simplest, if your father and grandfather died of a heart attack before the age of sixty, then your own life is at greater risk. Genetic predisposition to various disorders is an important factor when considering life expectancy, and life insurance companies take these factors into account when assessing whether an applicant is a safe risk. It is thought that conditions such as cancer, diabetes and heart disease are all influenced by genetic, as well as by environmental, factors.

If parents suspect their child has a genetic defect then, in some cases, the presence of the defective allele can be detected using a variety of genetic screening tests. For example, amniotic fluid containing embryo cells can be removed by **amniocentesis**. The cells of the embryo are then grown and examined for DNA and chromosome defects. Karyotypes can be examined for chromosome abnormalities, such as the one which causes Down's syndrome. Such tests carry an element of risk themselves, and so are not routinely carried out. Another disadvantage of amniocentesis is that it cannot be carried out until around the 16th week of pregnancy, after which the tests take a few weeks to perform. So, if a termination of the pregnancy is chosen, the embryo, which dies, is already well formed. Some people find this unacceptable and equate the action with murder.

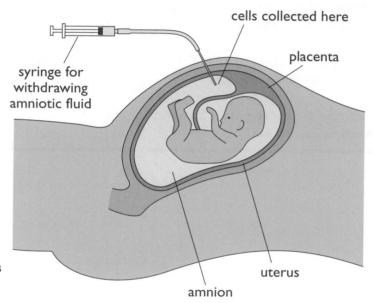

Amniocentesis

Another method of screening is to examine the blood cells or urine of a person. The disorder phenylketonuria (PKU) is checked for in this way (see *page 7*).

As genetic screening becomes more effective, and techniques in gene therapy improve, it is certain we will eventually be able to cure many genetic disorders. This has been made more possible by the recent revelation of the human **genome** – the entire genetic code for a human being. But these advances bring with them major moral and ethical issues which must also be addressed. For example:

- Where do we draw the line at genetic manipulation of the human genome? If we screen for cystic fibrosis, might we also screen for, and alter, a child's intelligence or even the colour of its hair?
- Should we grow then destroy human embryos to develop and test treatments for genetic disorders?
- As the genome of humans becomes easier to determine, should society allow anyone to buy a copy of his or her genome?

Questions

89 What term is used to describe the appearance of a person with respect to his/her genes?

90 A boy has a mother who has Huntington's chorea. What are the chances he will have inherited the HC gene from his mother?

91 What term is used to describe alleles which are equally dominant to one another?

92 A woman of blood group O has two children to a man of blood group AB. Which of the following are possible blood groups of their children?

   AB   A   B   O

93 A man and woman, whose fathers are both colour-blind, want to know the chances of having a child with colour-blindness. What would your advice be?

94 What term is used to describe the inheritance of characteristics affected by many genes?

95 What kind of mutation results in Down's syndrome?

96 A DNA code changes from AATCGATTC to AATCGATAC. What kind of mutation is this?

97 What kind of mutation is an inversion mutation?

98 What is a pathogen?

99 What term is used to describe the entire genetic complement of a person?

100 When might amniocentesis be carried out?

UNIT 2: THE CONTINUATION OF LIFE

> This 40 hour Unit contains information on reproduction, birth, growth and development, influence of hormones, circulation, delivery and removal of materials to and from cells, and on regulating mechanisms.

REPRODUCTION – FEMALE

Girls become sexually mature at **puberty**, at around the age of 10–14, and at that time their monthly **oestrous (menstrual)** cycle begins. This cycle will continue until the girl reaches the age of around 50, when it stops permanently. This 'change of life' is called the menopause. Normally, the only other occasion when the oestrous cycle stops is when a woman becomes pregnant. Then the cycle is held in check for the nine months of **gestation**, until the birth of the child. The cycle lasts for around four weeks and is controlled by changes in concentration of four hormones: **follicle-stimulating hormone (FSH)**, **luteinising hormone (LH)**, **oestrogen** and **progesterone**. The first two are manufactured in the **pituitary gland**, which is attached to the underside of the brain, and the other two are produced by structures in the ovaries.

The cycle is initiated by the production of FSH which has two effects:

- It causes an egg to mature and a ball of cells, called the **Graafian follicle**, to develop round the egg (ovum) in one ovary. A woman has thousands of eggs present in her ovaries from before birth, and during her life around 400 of these will mature to be released once a month.

- FSH also stimulates the follicle to produce oestrogen. Oestrogen in turn inhibits the production of FSH, so keeping its production in check. This is an example of **negative feedback control**. Negative feedback mechanisms are commonly employed by the body to control processes, or concentrations of various substances.

menstruation   uterus lining thickens ready to receive embryo   lining breaks down if no implantation takes place   menstruation

day 1   day 14   day 28

time

follicle maturing   ovulation   corpus luteum developing   corpus luteum breaks down

ovum

Oestrous cycle   oestrogen secreted   progesterone secreted

Oestrogen also has a number of other effects on the body; it initiates the repair of the uterine wall, in preparation for a developing embryo, and stimulates the pituitary gland to produce a sudden upsurge of LH. The surge of LH stimulates **ovulation** – the rupture of the follicle in the ovary and the release of the egg. LH also stimulates the development of the follicle into the **corpus luteum**, a small 'yellow body' in the ovary.

The corpus luteum produces progesterone for a few days before degenerating if the egg has not been fertilised. The progesterone which it produces during this short time encourages further development of the uterine lining in preparation for a new embryo. High concentrations of progesterone are maintained in the body if fertilisation occurs, due to the survival of the corpus luteum for a number of months. When it does finally disappear, the production of progesterone is taken over by the placenta until birth. The progesterone helps maintain the lining of the uterus and also, importantly, inhibits the pituitary gland and thereby holds ovulation in check. It would not make biological sense to have another fertilised egg arriving at the uterus when one was already there developing. This is one of the reasons progesterone is the principal ingredient in most contraceptive pills. Oestrogen too is often used to help keep the menstrual cycle in check. If fertilisation does not occur in a normal cycle, the levels of progesterone fall, allowing menstruation to take place.

REPRODUCTION – MALE

The manufacture of sperm occurs in the **seminiferous tubules** of the testes. The development of male characteristics is influenced and controlled by a variety of hormones, similar to those present in females.

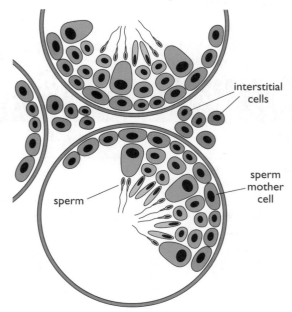

interstitial cells

sperm mother cell

sperm

Cross-section through seminiferous tubules

Testosterone is produced by the **interstitial cells** of the testes and influences the onset of puberty, the development of male sexual characteristics, and the development of sperm in the testes. Unlike most hormones, it is not a protein or peptide but a steroid (a lipid-like substance derived from cholesterol). Testosterone boosts basal metabolic rate and for this reason men tend to have a higher metabolic rate than women. The secretion of testosterone is influenced by two other hormones, **follicle-stimulating hormone (FSH)** and **luteinising hormone (LH)** which are produced by the pituitary gland. FSH, along with testosterone, indirectly stimulates the production of sperm in the testes. LH stimulates the interstitial cells to produce testosterone. In males, LH is therefore sometimes called, more logically, **interstitial cell-stimulating hormone (ICSH)**. Testosterone has a negative feedback effect on the pituitary gland and thereby concentrations of FSH, LH and testosterone are controlled.

During ejaculation, the sperm pass along the sperm duct and are joined by **seminal fluids** produced by the **prostate gland** and the **seminal vesicles**. These fluids stimulate the sperm to swim and provide them with sugar (fructose) as a source of energy.

Steroids

Anabolic steroids are synthetic lipids, similar in structure to testosterone. In the 1950s, athletes started to inject testosterone to improve muscle size. It was found that steroids also helped athletes extend their training period and it made them more aggressive, which gave them a competitive edge.

Steroids enter the nuclei of muscle cells and stimulate the cells to transcribe the genes for the manufacture of actin and myosin, hence increasing muscle size and strength. However, they have many harmful side-effects. Athletes were injecting up to 40 times the normal concentration and many developed symptoms such as acne, impotence, diabetes and heart disease. In addition, male athletes found that their testes decreased in size and produced fewer sperm. Female athletes found that their periods stopped, and they tended to develop male characteristics such as increased body hair and a deepening of the voice.

Today, athletes are banned from using these hormones, and other artificial substances, because they are unfair and dangerous.

INFERTILITY

Couples having regular intercourse at the fertile period (a few days either side of ovulation) can only expect a 30% chance of fertilisation in any month, and around 10% of couples are unable to conceive within a period of 2 years. The most common causes of infertility and their treatment are outlined in the table below:

TYPE OF INFERTILITY	POSSIBLE CAUSES	POSSIBLE TREATMENT
failure to ovulate	•hormone imbalance •ill health*, emotional stress	drugs which stimulate production of FSH and LH
blockage of uterine tubes	•infection •tissue growth (e.g. tumour)	surgery, *in vitro* **fertilisation**[†]
failure of implantation	•hormone imbalance	drugs which stimulate production of FSH and LH
low sperm counts	•hormone imbalance, •ill health*, emotional stress	hormone treatment, *in vitro* fertilisation, **artificial insemination**[‡]

* obesity, excessive alcohol consumption, excessive smoking, drugs and a variety of venereal diseases can all affect male and female fertility. Prolonged use of some contraceptives can also lead to permanent fertility problems.

[†] *in vitro* fertilisation is fertilisation in a glass dish. Sperm and eggs taken from the two potential parents are allowed to fuse 'in glass' in the laboratory. The zygote (fertilised egg) is allowed to divide a few times then replaced in the uterus of the woman to develop normally.

[‡] artificial insemination is the insertion of sperm (obtained by masturbation) into the vagina, using a mechanical device such as a syringe. The technique is used either when a man's sperm count is low or when his sperm are not viable (healthy), in which case another man's sperm are used.

CONTRACEPTION

Contraception is the prevention of fertilisation or implantation by some means or other. There are many different methods: some mechanical, such as the cap and condom; others chemical, such as spermicidal creams. Two 'biological' methods are described below:

FORM OF CONTRACEPTION	HOW IT WORKS	SUCCESS RATE
contraceptive pill – taken daily	contains progesterone, and sometimes oestrogen, which prevent ovulation	high
rhythm method – intercourse restricted to infertile period	infertile period is determined by checking temperature and vaginal mucus. Small rise in temperature (0·5°C) and thinning of vaginal mucus indicates ovulation	low

Questions
1 What term is used to describe the stage at which boys and girls become mature?
2 What is the length of gestation in humans?
3 Where are FSH and LH manufactured?
4 What two effects does FSH have on the ovary?
5 Oestrogen inhibits the production of FSH. What term describes this feedback effect?
6 Which hormone stimulates ovulation?
7 At what time of the monthly cycle is ovulation most likely?
8 What happens to the follicle after ovulation?
9 What hormone is produced by the corpus luteum?
10 Which two hormones stimulate the repair and maintenance of the uterine lining?
11 What effect do these two hormones have on the pituitary gland?
12 What happens to the corpus luteum if fertilisation does not occur?
13 What effect does a drop in progesterone have on the lining of the uterus?
14 Where precisely are sperm manufactured?
15 Name the hormone produced by the interstitial cells.
16 What kind of substance is testosterone?
17 Which two hormones interact together to stimulate the production of sperm?
18 What is the main function of ICSH?
19 Name two glands which produce seminal fluid.
20 When a woman is infertile because she cannot ovulate, what possible treatment might she receive?
21 What term is used to describe fertilisation in a test-tube outside the body of a woman?
22 Which two hormones are the common constituents of the contraceptive pill?
23 What changes take place in the vagina around the time of ovulation?

PRENATAL DEVELOPMENT

When a sperm fertilises an egg (in the oviduct), a **zygote** is formed. If two sperm fertilise two eggs, then **dizygotic** (fraternal) twins are formed. If one sperm fertilises one egg and the zygote later splits into two separate sets of cells, **monozygotic** (identical) twins are formed. After fertilisation, the zygote starts to divide as it continues on its way down the oviduct towards the womb. The first few divisions are called cleavage divisions because the cells simply split into two without any increase in size. As the cleavage divisions take place, cell size is continually halved until 'normal' cell size is achieved. The ball of cells arrives in the uterus a few days after fertilisation and eventually embeds itself in the wall (**endometrium**) of the uterus. This process is called **implantation** and usually occurs around one week after ovulation.

Shortly after implantation the embryonic cells start to specialise (**differentiate**) and form the organs of the body. By the end of the embryonic period, when the embryo is eight weeks old and 2cm long, all the adult organ systems are recognisable. After this point the embryo is referred to as a **fetus**. The fetus is genetically different from its mother and would be treated as a foreign invader were it not for the fact that the mother's immune system is specially adapted to cope with the presence of the fetus' tissues.

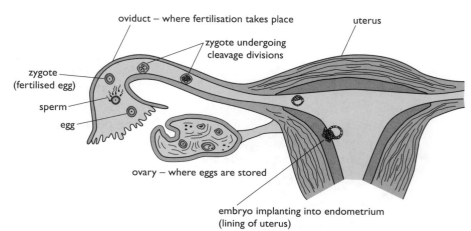

oviduct – where fertilisation takes place

zygote undergoing cleavage divisions

uterus

zygote (fertilised egg)

sperm

egg

ovary – where eggs are stored

embryo implanting into endometrium (lining of uterus)

Sometimes, however, the system breaks down. For example, if a Rhesus-negative (Rh⁻) woman has a Rhesus-positive (Rh⁺) child, there is a risk that she will produce antibodies against the Rhesus antigens on the red blood cells of her baby. This is not a problem with the first child because the response is too slow. But, if she has a second Rhesus-positive child, then there is a greater risk that antibodies will cross the placenta and destroy the child's blood cells. If a child is at risk, the standard treatment was 'exchange blood transfusion': Rh⁺ blood is removed from the baby and replaced with Rh⁻ blood. A more recent treatment is to inject the mother with anti-D factor which neutralises any Rh⁺ blood cells reaching the mother's blood. This prevents her from becoming sensitised and mounting an antibody attack.

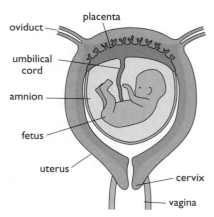

placenta

oviduct

umbilical cord

amnion

fetus

uterus

cervix

vagina

The fetus receives food and oxygen from its mother and gets rid of waste, such as urea, through the **placenta**.

In the placenta, the blood of the fetus and mother come into close contact but do not mix because the bloods may be of different blood groups, and because the mother's blood pressure would be too high for the fetus. Oxygen and carbon dioxide move across the placenta by **diffusion**, glucose by **active transport** and antibodies by **pinocytosis**. Unfortunately, many harmful substances can also pass from mother to fetus during this time. Alcohol, nicotine, many drugs, antibodies and viruses (e.g. rubella) can all pass across the placenta into the blood of the fetus and cause serious damage.

Oestrogen and progesterone are produced by the placenta towards the end of pregnancy and, as well as maintaining the endometrium, they stimulate the pituitary gland to produce a hormone, **prolactin**, which promotes the manufacture of milk by the breasts.

BIRTH

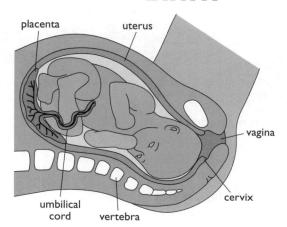

placenta   uterus

vagina

cervix

umbilical cord   vertebra

As the fetus increases in size towards the end of **gestation**, the uterus and cervix are stretched. This stretching process causes impulses to be sent to the **pituitary gland** where there is a release of the hormone **oxytocin**. As oxytocin concentration rises in the blood, it causes the contractions of 'labour'. Smooth muscle (involuntary muscle) in the walls of the uterus contracts in rhythmic waves, pushing the fetus out of the womb. The first sign that birth is about to happen is the flow of amniotic fluid from the vagina. Normally, a few minutes after the birth, the placenta is expelled.

Birth represents quite a shock to the infant. During the birth process, it is suddenly cast out of a warm watery environment into the cold, and its placental life support system is cut. Now the infant must do for itself all that the mother's body has been doing for it for the past nine months of gestation. It must breathe, feed, excrete and try to maintain its body temperature. Breathing is particularly difficult for premature babies who lack an important protein in their lungs which reduces the surface tension of water. Without this protein, the alveoli in the lungs collapse. Premature babies also lack fat stores.

Lactation is the production of milk by the breasts. For the first few days after birth, true milk is not produced. Instead, a yellowish fluid called **colostrum** is produced. Colostrum has a lower lactose content than milk and contains almost no fat, but it does have more protein, minerals and more vitamin A than true milk. Colostrum is particularly rich in antibodies which give the infant a degree of **passive immunity** against disease. However, breast milk can also contain poisons which the mother has unwittingly ingested in her food. Some persistent pesticides, such as DDT, which enter food chains can be detected in human breast milk (see *page 80*).

After birth, the suckling of the infant induces the secretion of the hormone **prolactin** by the pituitary gland. Each time it suckles, impulses travel from the nipple to the **hypothalamus** which stimulates the pituitary to produce prolactin and oxytocin. The effect of oxytocin is almost immediate, in that it stimulates the ejection of milk from the glands in the breasts. Oxytocin also causes the uterus to contract to normal size. The prolactin surge stimulates the breasts to manufacture more milk for the next feed, and also inhibits the activity of the pituitary gland in the production of FSH and LH. So, when a baby suckles it not only obtains a rich supply of food but also prepares the mother to provide for the next meal and ensures that no other competitors are going to come along in nine months' time to share this unique supply of nourishment. Suckling, and the oxytocin response, are examples of a positive feedback mechanism, because the more the child suckles, the more milk it gets. The hormone oxytocin is sometimes given to pregnant women to induce birth, to reduce bleeding after birth or to stimulate milk production.

Lactose intolerance

All mammals feed their young on milk, and humans are no exception. The main carbohydrate found in milk is lactose, a disaccharide. It would seem much simpler for milk to contain glucose, as it is easier to make and does not need to be digested. So why lactose? Glucose, a monosaccharide, is a very difficult substance to store, as it diffuses from one area to another very rapidly. This is one reason why we store glucose as glycogen in the body. It is also absorbed from the gut very rapidly. Lactose, on the other hand, is easier to store in one place and is slow to digest. So, it stays in the mother's milk, rather than diffusing to the surrounding tissue, and in the baby's gut, there is a steady, slow absorption of glucose over a period of time, rather than a sudden surge at feeding time.

Babies need the enzyme lactase to digest lactose, but many people lose the ability to make this enzyme as they get older, and this leads to an unpleasant condition called 'lactose intolerance'. The symptoms are pain in the gut, diarrhoea and flatulence. It is not common in northern Europeans, but is very common in people from Africa and Asia where, historically, milk products have not been a major part of the diet.

POSTNATAL DEVELOPMENT

Growth is most rapid before birth as a fetus can double its weight in a matter of weeks during early development. However, after birth, growth is still relatively rapid and declines slowly until it ceases altogether at around 20 years of age. There are variations though, the most marked of which is the growth spurt which takes place at **puberty**. This is earlier in girls than in boys and is marked not only by an increase in size but also by many other physical changes:

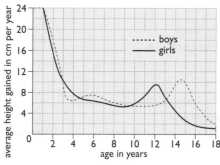

Average height increase per year

CHANGES IN FEMALES AT PUBERTY	CHANGES IN MALES AT PUBERTY
increased growth of hair on body	increased growth of hair on body and face
widening of hips	increased growth of skeleton and muscles
maturation of sex organs	maturation of sex organs
eggs start to mature – one per month	sperm are manufactured – 1000s per second
start of oestrous cycle	deepening of voice

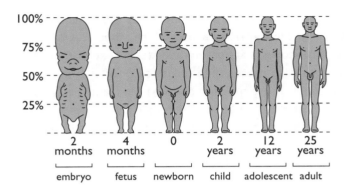

Growth progresses after birth and body proportions change, as shown in the diagram. At birth, the head is large, relative to the body, because the brain has to be well developed to cope with the large degree of learning which takes place during the early years.

Much of growth and development is controlled by a variety of hormones, the most important of which are given in the table below:

HORMONE	EFFECT OF HORMONE
growth hormone	promotes growth of bones and protein metabolism
thyroxine	controls cell growth and differentiation, and BMR*
oestrogen/progesterone and testosterone	promote the development of secondary sexual characteristics as described in the table above

* BMR = Basal Metabolic Rate. This is a measure of the basic chemical turnover of the body when at rest.

Questions

24 What is a zygote?

25 How many sperm and eggs fuse together to give dizygotic twins?

26 What are the first few divisions of the zygote called?

27 What is the endometrium?

28 Where does fertilisation take place?

29 When does implantation usually occur?

30 What is the name of the fluid in which the embryo develops?

31 What risk results to a Rhesus positive child being carried by a Rhesus negative mother?

32 How are gases such as oxygen and carbon dioxide carried across the placenta?

33 How is glucose carried across the placenta?

34 Name a molecule which is carried across the placenta by pinocytosis.

35 Name the hormones produced by the placenta towards the end of pregnancy.

36 Which hormone stimulates the production of milk?

37 What effect does oxytocin have on the muscles of the uterine wall?

38 What is the name of the first milk produced after the birth of the baby?

39 What effect does suckling have on the hypothalamus/pituitary gland?

40 What is the ratio of head length to body length of a newborn child?

41 Which hormone directly affects cell growth and BMR?

TRANSPORT SYSTEMS

All living things have a need to move substances around their bodies and the need for transport starts at cell level. All cells have to move substances round the cytoplasm and to and from their immediate environment. To do this they rely on a variety of methods, some of which have been studied already. In liquids and gases the ability of atoms and molecules to move on their own from areas of high concentration to areas of low concentration is called **diffusion** (see *page 21*). Our bodies make much use of this process, but the length of time it takes for a substance to diffuse from one place to another is proportional to the square of the distance. Put another way, if it takes one second for a molecule of glucose to diffuse across a cell, it will take over 50 years for the same molecule to diffuse a distance of one metre. Quite clearly, although diffusion is sufficiently quick to transport substances across cells, it is not sufficiently quick to get the glucose from the intestines to the active muscles of the arms and legs! So our bodies have developed other methods to increase the speed at which substances can be moved from one place to another. Peristalsis, breathing movements and the beating of the heart all aid the mass flow of materials round the body.

Another consequence of great size is the change in the relative size of surface area to volume of any object. A tiny cell has a huge surface area in relation to its volume, whereas we humans have a relatively small surface area compared to our volume. This can be shown by a simple piece of arithmetic in the table below.

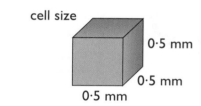

cell size

0·5 mm
0·5 mm
0·5 mm

The small cube represents a very large cell, and the large cube represents our own body volume. From the table it can be seen that moving from cell size to body size, the surface area : volume ratio drops one thousand fold. So, quite clearly, not only is diffusion too slow to allow substances to move from one end of the body to the other but also the surface area of our body must be increased considerably to allow substances to enter and leave it by diffusion. For this reason we have, for example, lungs with a surface area close to that of a tennis court to ensure adequate gas exchange.

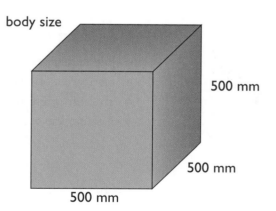

body size

500 mm
500 mm
500 mm
500 mm

DIMENSIONS	CUBE	
	side of 0·5 mm	*side of 500 mm*
Surface area (SA)	1·5 mm^2	1 500 000 mm^2
Volume	0·125 mm^3	125 000 000 mm^3
SA ÷ Volume	12	0·012

THE CIRCULATORY SYSTEM

The circulatory system consists of a four-chambered heart which pumps blood round the body. The blood travels through three types of blood vessel: arteries, veins and capillaries.

Arteries

thick muscular wall
(waterproof)

- have thick impermeable (waterproof) muscular walls
- pulsate due to the pumping action of the heart
- carry blood away from the heart to all the body organs
- have no valves
- contain oxygenated blood (except for the umbilical and pulmonary arteries)
- tend to lie deep in the body for protection
- divide into smaller arteries called arterioles

Veins

valve

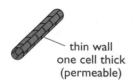

thin wall
(waterproof)

- have thin impermeable walls with very little muscle
- carry blood at low pressure towards the heart from the body organs
- do not pulsate
- have valves to prevent the back-flow of blood
- usually contain deoxygenated blood (except for umbilical and pulmonary veins)
- are not so well protected as arteries
- are divided into smaller veins called venules

Capillaries

thin wall
one cell thick
(permeable)

- join arterioles to venules
- have walls one cell thick
- are permeable to allow exchange of materials between blood and tissues
- have no valves

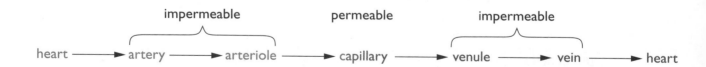

impermeable   permeable   impermeable

heart ——▶ artery ——▶ arteriole ——▶ capillary ——▶ venule ——▶ vein ——▶ heart

Blood consists of a fluid called the **plasma**, with red blood cells, white blood cells and platelets suspended in it. The composition of the blood is very precisely regulated by the lungs, liver and kidneys. The main arteries and veins of the body are shown in the diagram opposite.

The role of white blood cells – lymphocytes and macrophages – is outlined on page 51.

Platelets are tiny fragments of cytoplasm involved in the clotting action of blood.

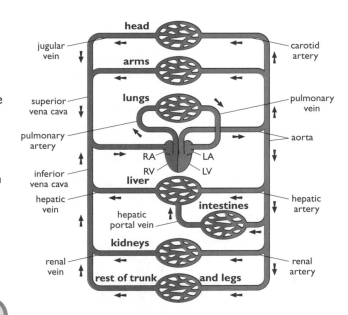

The main blood vessels of the body

Venous return

To say that valves prevent the backflow of blood is not quite the whole story. Actually, veins with valves represent hundreds of tiny 'hearts' throughout the body, all pumping blood back to the heart we all know. The veins themselves are passive – they don't beat, but they do assist in the return flow of blood to the heart.

When we walk, the bones and muscles in our legs move and bend and press the veins in our legs. Even when we breathe, we squeeze the large veins in our thorax. When a vein is squeezed the blood tries to flow away in both directions. But it can't, because of the valves, which only allow it to flow in one direction – towards the heart. So, as long as we are moving, the blood flow back to the heart is assisted by, what doctors call, venous return.

Of course, when you are sitting still in an aeroplane, this process does not operate and blood flow can slow to a point when clots can form. Then we suffer from deep-vein thrombosis (DVT).

Equally, when you are running, this process is very effective and assists in the rapid return of blood to the heart.

Questions

42 What happens to the surface area to volume ratio of a body as its size increases?

43 A cube has sides of 3cm in length. What is its surface area : volume ratio?

The answers to questions 44–49 are: artery, vein or capillary.

44 Which has the thickest walls?

45 On which one is pulse taken?

46 Which has valves?

47 Which has permeable walls?

48 Which has walls with layers of muscle?

49 Which takes blood to the heart?

50 Name the blood vessel which takes blood from:
    (i)   the heart to the brain
    (ii) the lungs to the heart
    (iii) the gut to the liver
    (iv) the kidney to the vena cava.

THE HEART AND THE CARDIAC CY

The average resting heart rate for adults is around 70 beats per min
although fit athletes have resting heart rates below 40 bpm. This is b
their heart chambers become larger and stronger as a result of regu
During extreme exercise, a young adult's heart rate can rise to over
The heart rate of a fetus and young baby is much higher than 70 bp
decreases as adult size is attained. It must be emphasised that these
figures and there is considerable variation from one person to anot
exercise, cigarette smoke and even the intake of food all have the ef
increasing heart rate.

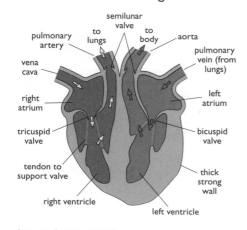

key: ⇨ deoxygenated blood
     ➡ oxygenated blood

The heart

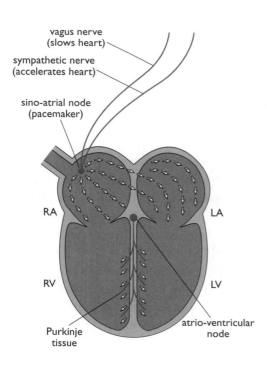

The heart has a unique form of muscle
can contract every 0·8 seconds for the
life of an individual. The fibres contain ac
myosin and are moderately striated. Ho
unlike striated muscle, the fibres are sho
usually uninucleate (one nucleus per fib
interconnected. The links between the f
them to contract at the same time. Abc
the volume of cardiac muscle fibres is t
large mitochondria. The cardiac muscle
on aerobic respiration almost exclusive
fatty acids most effectively, unlike other
the body which use glucose as a primar
coronary artery supplies the heart m
blood.

Most muscles contract as a result of in
reaching them from nerves. This is not
of cardiac muscle which will continue t
rhythmically even after its nerve supply
cut. In the wall of the right atrium is an
the **pacemaker**, or **sino-atrial node**
which generates rhythmic electrical wa
travel across the walls of the atria, mal
contract. Between the atria and the ve
another area called the **atrio-ventric**
(**AVN**) which transmits the impulse fr
to the ventricles so that they contract
after the atria. A network of fibres (Pu
spreads through the ventricles to carr
from the AVN to all the cardiac muscl
contracting phase of the cardiac cycle
systole and the relaxing phase is calle

pacemaker, if left on its own, would regulate the heart beat at around 80 beats per minute. However, the pacemaker is overridden by nerve lation from the **autonomic nervous system** via two nerves – the pathetic nerve accelerates heart rate and the **parasympathetic nerve agus nerve** as it is sometimes called) slows it down. In addition, heart s increased by the hormone **adrenaline**, produced by the **adrenal ds**, and the hormone **thyroxine**, produced by the **thyroid gland**. The omic control centre is found in the **medulla oblongata** of the brain o enable it to respond appropriately, it receives information from other rs in the body which detect blood pressure, temperature, oxygen and n dioxide concentrations.

heart beat can be measured by the pressure which spreads through the arteries and is called e. The heart can also be heard to beat using a oscope, the double 'lub-dub' sound arising from osing of the two atrioventricular (**bicuspid** and spid) valves followed by the two **semilunar** .

nportant technique to determine heart function measure the electrical changes which take place an **electrocardiograph** (**ECG**).

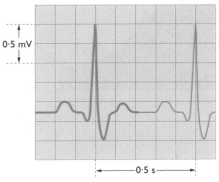

ECG trace taken during exercise

OOD PRESSURE

use blood flows in tubes of one kind or another, our circulatory system is o be a closed system. At any point in the system, blood will flow from 'A' so long as the pressure at 'A' is greater than the pressure at 'B'. It must , therefore, that blood pressure in the aorta is the highest, and blood ure in the vena cava the lowest, in the circulatory system. Pressure in the es fluctuates with the beating of the heart and is called **pulse**. Pulse and l pressure are measured on arteries and not veins.

d pressure is measured using a **sphygmomanometer**. This is an ment which measures two pressures, i.e. the maximum and minimum ures. The maximum is the pressure required to stop blood flowing down rachial artery in the arm – the **systolic pressure**. The minimum is ressure at which the blood is just able to flow freely along the brachial y – the **diastolic pressure**. The pressure is measured in millimetres of ury and is usually around 120/75 mmHg for a young adult. It tends to ase with age due to atherosclerosis (thickening of the artery lining), and ary considerably from one person to another, and from minute to minute ghout the day.

e blood flows through the arteries there is little drop in pressure because rteries have wide channels and elastic walls which offer little resistance e flow of blood. However, when the blood reaches the much smaller

arterioles, the pressure drops considerably. The blood pressure in the art[e
drops because there are many of them and they have a much smaller bor[e
the arteries. In effect this increases the surface area of the inner walls. It i[
friction of the blood fluid in contact with the walls which reduces the pre[s
and the flow of the blood.

By the time the blood reaches the capillaries,
flow rate and pressure have dropped consider
This is a desirable outcome because low pres[s
reduces fluid loss through the capillary walls a[
low flow rate allows time for exchange of ma[
to take place between the plasma and the tiss[

The blood pressure in the pulmonary circulat[i
five times lower than that in the systemic (ge[
body) circulation. Since the blood does not ha[
far to travel through the lungs this is not surp[
However, there is another important reason f[
low pressure. The blood in the lungs must tra[
through delicate thin-walled capillaries in intim[
contact with the air. If pressure were too high[
would leak out of the capillaries and fill the al[

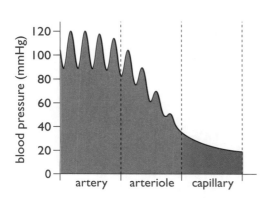

**Blood pressure in vessels supplying
muscles at rest**

Questions

51 In the example of an ECG trace shown on page 49, what is the heart rate of the individual shown?

52 A man's heart beats once every 0·8 seconds. What is his heart rate per minute?

53 Which artery supplies the heart muscle with oxygenated blood?

54 Which side of the heart deals with oxygenated blood?

55 Name the valve which separates the right ventricle and right atrium.

56 Where is the pacemaker found in the heart?

57 Give another name for the pacemaker.

58 Which branch of the nervous system controls heart rate?

59 What is the effect of stimulation of the sympathetic nerve on heart rate?

60 What part of the brain is involved in the control of heart rate?

61 What terms are used to describe the contracting and relaxing phases of the heart beat?

62 Why is blood pressure given as two values?

63 In which blood vessel is blood pressure at its lowest?

64 What changes to the arteries of older people causes an increase in blood pressure?

PLASMA, TISSUE FLUID AND LYMPH

Because the blood pressure in the capillaries is higher than in the surrounding tissues, some blood **plasma** fluid does leak through the permeable capillary walls. This is absolutely essential of course, because all the cells of the body need a supply of water, nutrients and oxygen. The blood plasma which leaks into the tissues is then called, not surprisingly, **tissue fluid**. Although some of the tissue fluid returns to the blood at the venous end of the capillary system due to osmosis, a proportion does not return to the blood. This lost fluid is collected by the **lymph system** which is found in almost every tissue except for bone, teeth and the central nervous system (CNS). Every 24 hours, around three litres of the tissue fluid is collected by the tiny blind-ended capillaries called **lymphatics**. Pressure in the lymphatics is even lower than pressure in the tissue fluid, so tissue fluid gradually flows into the lymphatics and moves along the lymph vessels to return to the main circulation at a point very near the heart.

Lymph vessels, like veins, have valves which help maintain the flow of fluid so long as the body is moving. This tissue fluid in the lymph vessels is now called **lymph**. As it flows back towards the heart it flows through many **lymph glands (lymph nodes)** where harmful bacteria are removed or inactivated by the action of **macrophages** and **lymphocytes**, Remember that lymphocytes originate in the bone marrow but migrate to the lymph glands where they divide further and become active. Lymph glands swell up during times of illness (e.g. tonsillitis) as a consequence of their increased activity. The spleen and tonsils are part of the lymph system.

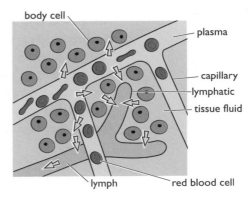

Relationship between lymph vessels and capillaries

As a consequence of the flow of fluid in the circulatory system described above, plasma, tissue fluid and lymph differ in their composition. Most plasma proteins are too large to pass through the walls of the capillaries, so there is much less protein in tissue fluid than in the plasma. For the same reason, red blood cells cannot escape from capillaries into the tissues but white blood cells are capable of squeezing between gaps in the capillary walls. Also, lymph tends to have more fat suspended in it because it is the specialised lymphatics of the villi, called **lacteals**, which collect digested fat and deliver it to the main circulation (see *page 54*).

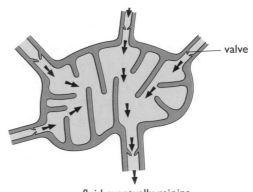

fluid eventually rejoins
blood near the heart

Lymph node

RED BLOOD CELLS

Red blood cells are not true cells, in so far as they do not contain a nucleus and have few organelles in their cytoplasm. Mitochondria are missing because the principal role of red blood cells is to transport oxygen. It would be inefficient to have mitochondria using up the oxygen en route from the lungs to the respiring tissues of the body.

Red blood cells are very small and biconcave in shape. Both these features have the effect of increasing the surface area of the cells relative to their volume. This means that oxygen can enter and leave the cells readily. What is more, oxygen does not have far to travel once inside the cell.

Mature red blood cells are full of haemoglobin, a red protein which has a high attraction (affinity) for oxygen. The small size of the corpuscles and the flexibility of their membranes also allow them to squeeze through the narrowest of capillaries.

Red blood cells are manufactured in the bone marrow. **Iron, vitamin B$_{12}$** and **amino acids** are the principal requirements for their manufacture. As they have no nucleus and no ribosomes, they are unable to regenerate worn out proteins and so can only live for around 120 days. After their short life, the cells are destroyed in the liver and much of the iron is recycled. The haemoglobin is degraded to **bilirubin** which is then excreted in the **bile**. Anyone suffering from a shortage of red blood cells or haemoglobin is described as anaemic.

←— 10 μm —→

Red blood cell

GAS TRANSPORT

The total pressure exerted by a mixture of gases is the sum of the pressures exerted independently by each gas in the mixture. The pressure exerted by each gas is called its **partial pressure** and is directly proportional to its percentage in the total gas mixture.

Since atmospheric pressure will support around 760 mmHg (millimetres of mercury) and, since oxygen occupies around 20% of fresh air, then the partial pressure of oxygen in fresh air is just over 150 mmHg. In the lungs there is always a lower percentage of oxygen than there is in fresh air. This is because inflowing air is mixed with stale residual air which is low in oxygen. So the partial pressure of alveolar air can be as low as 100 mmHg. However, this low figure presents no problem to haemoglobin which can become almost 100% saturated with oxygen at a partial pressure as low as 90 mmHg. When the haemoglobin reaches the respiring tissues, the partial pressure of oxygen is much lower there because the tissues are using it up. In these circumstances the haemoglobin is not able to keep all its oxygen and unloads about 30% of its oxygen very readily. What is more, if the temperature rises, or if pH drops, the haemoglobin will release even more oxygen. This can be seen in the graph shown opposite where the **oxygen dissociation** curve shifts to the *right* at higher temperatures. So, when the muscles are very active and generating much heat, the haemoglobin responds by releasing its oxygen more readily.

The loading and unloading of oxygen in the body normally takes place in the region where the dissociation curve is at its steepest. For this reason, the haemoglobin is very efficient at picking up oxygen in the lungs and releasing it to the respiring cells.

As well as supplying the body with oxygen, the lungs are also involved in the removal of carbon dioxide. The carbon dioxide, a by-product of respiration, is carried by the blood in three different ways: dissolved in the plasma (7%), as bicarbonate ions in the plasma (73%) and in the red blood cells (20%).

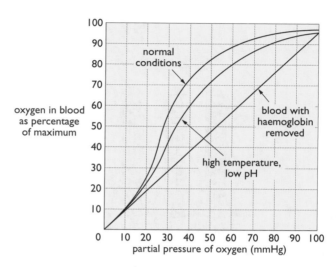

Blood which passes through respiring tissues picks up carbon dioxide and carries it to the lungs where the carbon dioxide diffuses from the blood into the alveoli. High concentrations of carbon dioxide in the blood stimulate the breathing centres of the brain to cause an increase in breathing rate.

As cell respiration speeds up to produce more ATP for exercising muscles, the increased concentration of carbon dioxide causes the breathing to quicken and deepen and at the same time helps the haemoglobin carry more oxygen to the muscles. Strangely, perhaps, changes in the oxygen content of the blood have a limited effect on breathing rates.

Questions

65 What is blood plasma?

66 What is blood plasma called once it has leaked out of the capillaries?

67 Which circulatory system collects the leaked fluid and returns it to the blood?

68 What happens to lymph fluid as it passes through the lymph glands?

69 Which blood cells are particularly active in the lymph glands?

70 What is the major difference between the composition of plasma and lymph?

71 What major organelles are missing from red blood cells?

72 What is the name of the protein which gives blood its colour?

73 Give one reason why blood cells are very small.

74 What is the average life span of a red blood cell?

75 Which vitamin is particularly important in the manufacture of red blood cells?

76 Where are red blood cells destroyed?

77 What is the fate of the haemoglobin during this process?

78 What effect does high temperature have on the ability of haemoglobin to release its oxygen?

79 What name is given to the graph on this page?

80 What proportion of carbon dioxide released by the respiring cells is carried by red blood cells?

81 What effect does a high concentration of carbon dioxide have on breathing rate?

THE ABSORPTION OF NUTRIENTS

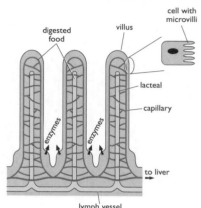

Villi of the gut

The small intestine is around 6 metres long and much of its inner lining is folded and covered with millions of tiny projections called **villi**. These in turn have a covering of epithelial (lining) cells which themselves have many microscopic projections called microvilli. These microvilli are the product of microscopic folding of the plasma membrane of each cell. This combination of folding at three different levels of magnitude gives the small intestine a very high surface area (about $600m^2$) which is needed for the absorption of digested food.

Each villus contains many blood capillaries which carry away the digested food. In the centre of each villus lies a branch of the lymph system, a tiny blind-ended tube called a **lacteal**, which carries away most of the digested fats. These fats eventually join the blood at a point near the heart where the lymph system empties its contents into the bloodstream.

Endocytosis, exocytosis, active transport and diffusion all play a part in the transfer of digested foods through the epithelial cells and into the capillaries and the lacteal. Some foods are so easily absorbed that the process starts as soon as they enter the mouth. For example, salt, glucose, alcohol and aspirin can all be absorbed in the mouth. Other foodstuffs require much assistance in crossing the intestinal barrier. For example, **vitamin B$_{12}$** needs the presence of a chemical called **intrinsic factor**. The vitamin combines with the intrinsic factor which then stimulates special cells to absorb it by endocytosis. Vitamins A, D, K and E need fats to transport them by diffusion into the blood.

THE LIVER

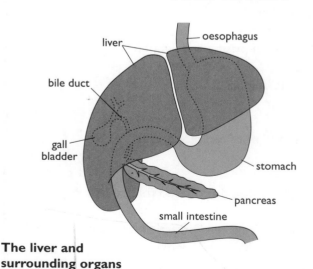

The liver and surrounding organs

The liver is the largest organ of the body and, because of its great metabolic activity, produces more heat than any other organ. The functions of the liver are many, but its prime purpose is to maintain the steady concentration of substances in the blood, i.e. it is a **homeostatic** organ. For example, it is important that the concentration of sugar in the blood is kept fairly steady at around one gram per litre. Our intake of sugar varies considerably through the day. So it is the liver's function to even out these variations; to remove sugar from the blood when there is too much, and to release it into the blood when there is too little.

A complete list of the functions of the liver would take many pages, but a number of important functions are given below:

- Excess amino acids are converted to keto acids and ammonia in a process called **deamination**. The body cannot store proteins, so this function is essential. The ammonia is very toxic and is converted immediately to urea which is removed from the blood by the kidneys. The keto acids enter the respiratory pathways and can be stored as glycogen or oxidised to produce ATP.

- Some amino acids can be made from others by a process called transamination. Amino acids which can be made in this way are called 'non-essential' because they are not absolutely essential in the diet.

- Plasma proteins are manufactured by the liver, including those involved in the clotting action of the blood.

- Old red blood cells are removed by phagocytic cells. The iron is stored and transferred to the bone marrow to be used again. Some of the breakdown products of haemoglobin are excreted in the bile as **bilirubin**, giving it a green-yellow colour. The brown colour of faeces and the yellow colour of urine both derive from the breakdown products of bilirubin.

- **Bile** is produced by the liver and stored in the **gall bladder** to be released into the duodenum (first part of small intestine) from time to time to aid digestion. Bile is an alkaline fluid containing **bile salts** which emulsify (break up into tiny droplets) fats. Bile is an activator for the enzyme lipase but does not itself contain any enzymes.

- The fat-soluble vitamins A, D and B_{12} are stored in the liver.

- Cholesterol, a lipid, is manufactured from acetyl coenzyme A (acetyl CoA). Although excess cholesterol is thought to contribute to arterial disorders, cholesterol is in fact essential for the manufacture of steroid hormones and vitamin D, and is a major component of membranes.

- Glucose and glycogen can be interconverted under the influence of a variety of hormones (see *page 57*).

- Poisons (**toxins**), such as alcohol and many drugs, are detoxified and removed from the blood.

- Hormones are converted to inactive compounds in the liver, so limiting their period of activity within the body.

The blood supply to the liver is unique. Unlike any other organ in the body, it is supplied by blood from an artery and a vein. The **hepatic portal vein** brings blood from the intestines, carrying the digested food which has been absorbed. Since the intestines have used up much of the oxygen, the liver gets its own separate supply of oxygen directly from the aorta via the **hepatic artery**. After blood has been dealt with by the liver it rejoins the main circulation through the **hepatic vein** (see *page 47*).

Questions

82 What is the function of villi?

83 What name is given to the central lymph vessel in a villus?

84 Which foodstuffs are transported by lacteals?

85 What substance is required for the proper absorption of vitamin B_{12}?

86 What is the name of the process in which amino acids are broken down in the liver?

87 What is the toxic end-product of this process?

88 Where is bile stored?

89 What is the prime function of bile?

90 Which vitamins can be stored in the liver?

THE KIDNEYS

We have two kidneys and their function is to regulate the fluid balance of the body (**osmoregulate**) and to remove nitrogenous waste such as **urea** from the blood. Approximately one thousand litres of blood pass through each kidney every day.

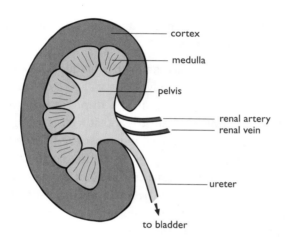

The process at first may seem rather unusual and inefficient. Each kidney contains around one million filtering units called **nephrons**. Around 180 litres of fluid per day are allowed to escape through tiny coiled knots of capillaries in the nephrons called **glomeruli** and then 99% of this fluid is **reabsorbed**. This requires an enormous amount of energy and the kidneys consume 10% of all the oxygen used by the body at rest. It might seem far more efficient to select poisonous molecules from the blood and only remove them. However, this would be a hazardous process for cells as they would then have to deal with high concentrations of poisonous materials. By allowing all the fluid to pass through the Bowman's capsule and then only reabsorbing harmless molecules, substances such as urea are dealt with indirectly in very dilute quantities.

The glomeruli have tiny pores which act as micro-filters, allowing almost all the blood plasma to pass through except for cells and large plasma protein molecules. The re-absorption process takes place in a tiny coiled tubule (**proximal convoluted tubule**) attached to the **Bowman's capsule**. All useful substances such as glucose, vitamins, and amino acids are reabsorbed along with much of the water and salts, leaving only enough water to dissolve the urea and excess salt for excretion. This first segment of the tubule is lined with highly active cells covered in microvilli (see *page 54*).

Further along the tiny tubule, at the **loop of Henlé**, salts are pumped into the medulla of the kidney to increase the osmotic concentration of the tissue fluid there. When water is in short supply, the pituitary gland secretes a hormone called **anti-diuretic hormone (ADH)** which makes the walls of the **collecting ducts** more permeable to water. So water passes by osmosis from the filtrate back to the blood to make it more dilute, and, as a consequence, the urine becomes more concentrated.

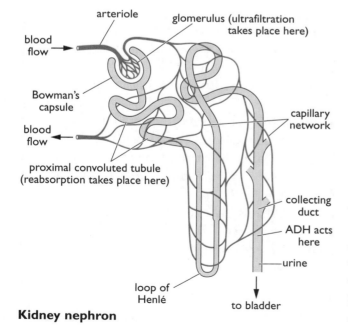

Kidney nephron

REGULATING MECHANISMS

Cells and their enzymes are very sensitive to changes in their surroundings. For this reason the body maintains the composition of body fluids and its temperature within very narrow limits. The process is called **homeostasis** and involves sensors of several kinds and a method of control called **negative feedback**. By this process, when any factor deviates from its set point, various correction mechanisms come into play to correct the deviation. When the deviation has been corrected, these mechanisms cease to operate until another change is detected.

The control of blood sugar

The concentration of sugar in the blood is held at just under one gram per litre, principally by the activities of the liver and pancreas. The hormone **glucagon** stimulates the release of glucose from glycogen stores. The hormone **insulin** has the opposite effect. Insulin and glucagon are manufactured in the **Islets of Langerhans** in the **pancreas** and it is these cells which detect changes in glucose concentrations in the blood. If glucose levels drop, glucagon is secreted, glucose levels return to normal and the stimulus for glucagon secretion disappears. If glucose levels rise, insulin is secreted, glucose levels return to normal and the stimulus for insulin secretion disappears.
These are examples of **negative feedback responses**. Other hormones, such as adrenaline and thyroxine, also influence blood sugar concentrations.

The control of temperature

The temperature of the blood is monitored by the **hypothalamus** at the base of the brain. When body temperature rises much above 37°C, various mechanisms are brought into play to reduce temperature. Sweating results in the evaporation of water from the skin and this takes heat energy from the body. The widening of arterioles (**vasodilation**), to allow more blood to flow near the surface of the skin, helps heat to escape more easily. Conversely, shivering and **vasoconstriction** help generate and conserve heat respectively. The raising of hair traps air which is a good insulator. In the long term, the body's metabolic rate can also change to generate more, or less, heat. These mechanisms, however, are **involuntary** (not under conscious control). We can consciously help to maintain a steady temperature, for example, by seeking shelter, changing our clothes, curling up in a ball or stretching out.

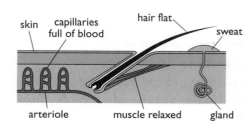

Hot

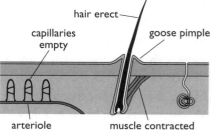

Cold

HYPOTHERMIA

Hypothermia sets in when the temperature corrective mechanisms fail to maintain body temperature and core temperature falls below 35°C. Watersports enthusiasts and mountaineers are particularly vulnerable to hypothermia, as are the very old and the very young.

Wet clothing can conduct heat from the body ten times more quickly than dry clothing, particularly when it is windy. This is because water draws heat from its surroundings as it changes from liquid to vapour. Water is also a better conductor of heat than air, so human survival time when immersed in water at 5°C can be less than one hour.

Young children are particularly at risk for a number of reasons. Being small, they have a large surface area : volume ratio. Since heat is generated in the 'volume' of the child and lost through the 'surface area', there is a relatively small source of heat-generating capacity and a relatively large area over which any heat generated is lost. Infants also have underdeveloped automatic temperature regulation mechanisms and have no **voluntary** responses, i.e. they cannot take conscious actions, such as putting extra clothes on, or seeking shelter, to remedy the situation.

Old people are particularly at risk because their metabolic rates decrease with age and their temperature regulation mechanisms are not so effective. For example, their vasoconstriction response tends to be weak, and their muscles are less able to generate heat energy during shivering. In addition, old people tend to be less active, they may have a poor diet and they may be reluctant or unable to warm their houses effectively in winter for financial or other reasons.

Questions

91 What are the microscopic filtering units of the kidneys called?
92 Name the tube which carries urine from the kidney to the bladder.
93 In what part of the nephron does filtration take place?
94 In what part of the nephron does most of the reabsorption take place?
95 Name one important substance which is completely reabsorbed by the kidney.
96 What is the name of the hormone which influences the reabsorption of water?
97 Where is this hormone produced and when?
98 What is meant by the term 'homeostasis'?
99 Which two hormones are manufactured by the Islets of Langerhans?
100 Which of these hormones is produced during times of hunger?
101 Name the storage carbohydrate found in the liver and muscles.
102 What part of the brain monitors body temperature?
103 What happens to skin arterioles when we are too hot?
104 What must happen to sweat before it has a cooling effect?
105 Which two groups of people are particularly at risk from hypothermia?

UNIT 3: BEHAVIOUR, POPULATION AND THE ENVIRONMENT

> This 40 hour Unit contains information on the nervous system, the brain, memory, human behaviour, the rise in the human population and humans' impact on the environment.

THE BRAIN

The brain is composed of some 10^{11} **neurones** (nerve cells). Each neurone can be linked directly to hundreds or even thousands of other neurones, so the potential for almost infinite complexity is vast. The human brain is not the largest of the animal world but, compared to human body size, it is relatively enormous. It is the computing power of the human brain which has allowed us to become conscious animals able to communicate in a sophisticated way and to engineer our environment to our liking in a way unparalleled by any other species on the planet. Surprisingly, although the human brain is the most important organ of the body, and the most significant in terms of the difference between ourselves and other mammals, relatively little is known about its functioning.

The largest part of the brain is the **cerebrum**, whose **cortex** (surface) is heavily **convoluted** (folded). The folding allows an increased number of cell bodies (50 billion of them) to be located in the cortex, so maximising the potential for interconnections between neurones. In fact, the folding is so considerable that the cortex, if laid flat, would have approximately the same area as a double page of a broadsheet newspaper. From these cell bodies, a mass of **axons** (fibres) runs under the cortex to connect them to one another, to other cell masses in the brain, and to the rest of the body by way of the spinal cord. The cerebrum is divided into two halves: the two **cerebral hemispheres**, which are quite separate from one another apart from a mass of fibres, called the **corpus callosum**, which transfers information from one hemisphere to the other.

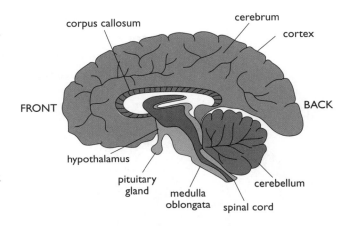

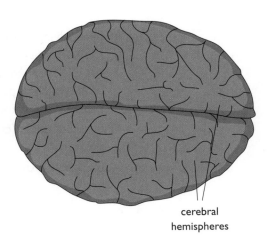

The brain

Attached to the underside of the brain at the back is the **cerebellum** which is concerned with the unconscious fine control of voluntary muscle movements and balance. The whole brain sits on the brain stem which connects the brain via the **medulla oblongata** to the spinal cord. The medulla is concerned with the unconscious co-ordination of basic functions: breathing, heart rate, digestion and **reflex actions** such as sneezing, coughing and vomiting.

Despite its great complexity, the brain is very flexible. For example, after injury, the functions of the damaged part can sometimes be taken over by unaffected parts. This flexibility of the brain is described as its **plasticity**.

The brain receives information from sensory receptors, for example, the eyes, the ears and the skin. It then deals with this information and may act on it by sending out information along motor neurones to muscles or glands. The **sensory** and **motor** areas of the brain are fairly clearly defined but many higher mental functions, such as memory, language, consciousness and emotion appear to have overlapping domains and are not nearly so easy to pinpoint.

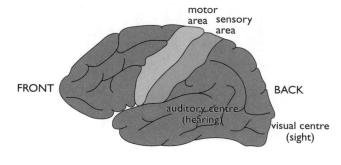

The diagram shows the cerebral cortex and the centres of control. Not surprisingly, the size of any segment of the motor or sensory areas of the brain is directly related to the function which that area carries out. This is best illustrated by the diagram of the human shown below which has each body part drawn in proportion to the area of the cerebral cortex devoted to its motor control.

The functioning of the brain has been studied in a variety of ways:

- by studying the effect of brain damage (split brain, stroke or other injury) on the behaviour of individuals. An individual whose corpus callosum has been cut is described as 'split-brain'. Strokes result from blocked arteries in the brain which deprive cells of oxygen. Brain cells have a high metabolic activity and can only survive for a few minutes without oxygen.
- by using EEGs (electroencephalograms) which record the brain's electrical activity.
- by carrying out brain scans to highlight areas of high metabolic activity in the brain.

The brain and the spinal cord together make up the **central nervous system** (**CNS**). This is the processing system. Nerves from the CNS radiate to all parts of the body and make up what is called the **peripheral nervous system** (**PNS**). The PNS simply links the CNS to the world around us and is divided into the **somatic** and **autonomic** nervous systems. The somatic system is used for all conscious activities such as speaking, writing and walking.

The autonomic nervous system controls automatically our basic body functions such as heart rate, breathing, blood pressure and perspiration (sweating). The autonomic system has two branches which work **antagonistically** (in opposition):

● the **sympathetic system** prepares the body for action
● the **parasympathetic system** encourages the conservation of bodily resources.

So the sympathetic system diverts blood from our digestive system to muscles during times of stress or exercise. It also stimulates sweating, breathing, heart rate and dilation of the pupils. The parasympathetic system has the opposite effect and comes into play during times of relaxation and rest.

NEURONES

Neurones (nerve cells) conduct electrical impulses from one part of the body to another. They consist of a **cell body** with numerous cytoplasmic extensions called **fibres**. The cell body contains all the usual organelles except for centrioles. This means that it cannot divide and needs to function throughout the life of its owner. Unfortunately, if neurones are damaged, regeneration is very poor. Hence paralysis after an accident is usually for life.

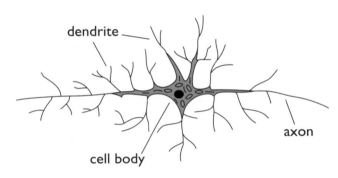

The fibres are of two types: **axons** and **dendrites**. Axons can be exceedingly long (up to a metre in length) and they carry impulses away from the cell body. Dendrites carry impulses towards the cell body. The axons are often covered with an insulating **myelin sheath** which increases the speed of impulse passing along the neurone. Also, the larger the diameter of the axon, the greater the speed of impulse. The myelin sheath is formed by Schwann cells which wind their membrane round the axon many times. This process is not complete at birth and this is one reason why the muscular movements of infants are less co-ordinated. Multiple sclerosis is a disorder in which the myelin sheaths of nerve fibres degenerate, so affected individuals slowly lose the ability to control their muscles.

There are basically three different types of neurone:

● **sensory neurones** carry information from the senses to the brain
● **motor neurones** carry instructions from the brain to the muscles and glands of the body
● **association (relay) neurones** are confined to the CNS and link with one another and with sensory and motor neurones.

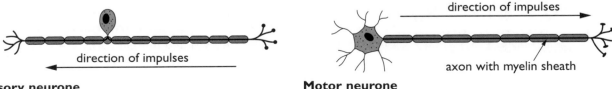

Sensory neurone   **Motor neurone**

Bundles of sensory and motor fibres linking parts of the body with the brain are called **nerves**.

Neurones are linked to one another in an almost infinitely complex way but two types of neural pathway can be easily understood: **converging** and **diverging** pathways. In converging pathways, two or more fibres meet at one neurone. In the **retina** of the eye, light-sensitive cells, called **rods**, detect very weak light and pass on a weak signal towards the brain. By bringing these signals together along converging pathways, the enhanced effect is sufficiently strong to be detected by the brain. Sometimes, conflicting signals converge at a junction. In such cases one or other is dominant and stimulates or inhibits the next neurone. This can be tested with **reflex actions** such as blinking where one can try to prevent the automatic reflex by conscious will. Some people find this easy, whereas others do not.

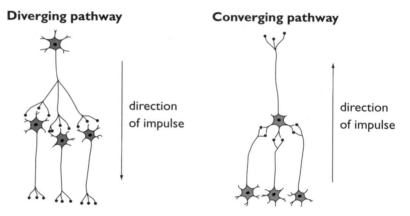

Diverging pathways allow one signal from the brain to be sent to many destinations simultaneously. For example, when catching a ball, all the muscles of the arm which have to be contracted or relaxed at the correct moment can do so in synchronisation.

NEUROTRANSMITTERS

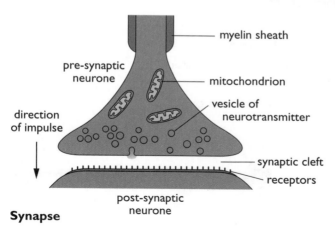

Synapse

The point where two nerve fibres meet is called a **synapse**. At the synapse the nerve fibres do not touch. Instead there is a tiny gap (synaptic cleft) of only around 20nm (one nm = one millionth of a millimetre) across which chemicals must diffuse to pass the impulse to the next neurone. Such chemicals are called **neurotransmitters** and there are many different kinds, the best known of which are **acetylcholine** and **noradrenaline**. Junctions between nerves and muscle fibres are very similar to synapses and are known as **neuromuscular junctions.**

When an electrical impulse reaches the synapse, it causes vesicles in the pre-synaptic axon to release a transmitter chemical which diffuses across the gap to receptors on the other side. There are many different kinds of **receptors** and they are affected in different ways by the same neurotransmitter. Incoming neurotransmitters excite some receptors and inhibit others. If insufficient neurotransmitter is produced, the stimulus is not passed on to the next neurone.

The neurotransmitters must be removed after they have acted, otherwise they would continue to have their effect. Acetylcholine, which stimulates the activity of skeletal muscle, is removed by enzyme action. Noradrenaline is reabsorbed by the pre-synaptic neurone. The continual synthesis and removal of neurotransmitters requires an enormous amount of energy and, as a consequence, neurones have a very high metabolic rate. It is for this reason that the brain is so easily damaged as a result of even just a few minutes' lack of oxygen. It is worth noting that behaviour-altering drugs (such as nicotine, alcohol, cocaine, ecstasy and heroin, and anaesthetics used in hospitals) produce their effects by influencing synaptic transmission.

Questions

1   What is a neurone?
2   Which part of the brain is largest in humans?
3   Which part of the brain controls heart rate and breathing?
4   What does the cerebellum control?
5   What name is given to the band of fibres which links the two cerebral hemispheres?
6   In which direction do impulses flow through axons – away from or towards the cell body?
7   Name the sheath which protects each axon.
8   What effect does this sheath have on the speed of transmission of nervous impulses?
9   Which part of the brain controls reflex actions?
10   In what part of the brain are the sensory and motor strips found?
11   What two components make up the central nervous system?
12   Name the branch of the nervous system which deals with all conscious activities.
13   Which branch of the autonomic nervous system prepares the body for action?
14   Name the three principal types of neurone.
15   What name is given to a nervous pathway where one signal splits into many?
16   Where are rods found, and what is their function?
17   What is a synapse?
18   How do nervous impulses cross a synapse?
19   Give an example of a neurotransmitter.
20   What happens to neurotransmitters after they have had their effect?

MEMORY

When we think of memory, we are inclined to think of our own ability to remember facts for examinations or of our memory of distant events (usually pleasant ones). However, our ability to remember is of far greater importance. For example, as you read this page, if you were unable to remember the words from the previous line then the process of reading would become meaningless. Even listening to speech requires use of memory. Some of us claim to have poor memories but we can all speak at least one language, read and write, and remember faces and facts from years gone by with little difficulty. We all have excellent memories and it is our great power to memorise language and symbols which makes us uniquely human. The brain has various areas involved in memory, one of which is the **limbic system**.

The limbic system

Around the central core of the brain are a number of areas which together make up the limbic system. The limbic system is closely connected to the hypothalamus and seems to be involved in various aspects of memory and personality. If parts of the limbic system have to be removed by surgery, memory can be affected such that memory of distant events remains normal but memory of recent events, within the previous year, seems to have disappeared. Moreover, any subsequent happenings are not remembered. The neurones of the limbic system are rich in one particular receptor known as **NMDA** (N-methyl D-aspartic acid). In **Alzheimer's disease**, where memory loss is the main symptom, there appears to be a loss of the ability to make acetylcholine in the cells of the limbic system. These facts suggest that NMDA and acetylcholine are involved in the process of memory storage.

There are three stages in memorising facts:

ENCODING   >   STORAGE   >   RETRIEVAL

In simple terms, when we memorise something, we turn it into a form which can be retained in our memory; we keep it there for a period of time which can be anything from a few seconds to over 100 years, and we access it from memory for use when required. The ease with which we carry out each of these functions varies enormously and depends on many other factors. For example, when we fail to remember someone's name this can be due to a failure of any one of these stages. It seems likely too that we have different forms of storage for different functions. For example, the ability to ride a bike is learned and never forgotten, but it seems that this form of memory is different from the one in which we memorise our friends' names.

What is more, our memory of words as they are spoken operates in a different way to our memory of, say, our address. When we listen or read, we use our **short term memory (STM)** to remember each word for a few seconds to enable us to understand each sentence. When we try to remember something for longer than just a few seconds we use our **long term memory (LTM)**.

Short term memory has a very limited capacity. On average, the capacity seems to be around 7 items ± 2. This is called the **memory span**. It is very easy to test memory span. Simply read out unrelated lists of numbers or letters to another person and test how many they can recall correctly.

Any list longer than 8 items becomes almost impossible to remember without using other specialised techniques involving LTM.

The majority of information encoded in memory is encoded as sounds (**acoustic coding**) but we can also encode information as pictures (**visual coding**) or by using meaning (**semantic coding**). Information can be held in short term memory for about 30 seconds. For example, when you look up a telephone directory, you often remember the phone number just long enough to use it – it has been stored in STM and not transferred to LTM. You can remember the number in STM so long as it contains around seven **chunks** (bits) or less, e.g. 842561. Most people remember a number or a series of letters like this by **rehearsing** it briefly, i.e. by saying it over in their heads a few times. This is **acoustic** coding. Visual coding and semantic coding can also be used to store memories. If we used visual coding to remember the telephone number, we would have a mental picture of the number in our mind's eye. If we used semantic coding we would place meaning round the numbers and remember the meaning rather than the numbers.

This technique enables us to remember more than seven numbers by **chunking** them. If the telephone number was 0141 654 1352 we might remember the first four digits as one chunk because our LTM knows that 0141 is the code for Glasgow, 654 might be remembered more easily because the digits decrease by one, and 1352 might be remembered as one item if we recognise that 13 is the number of playing cards in a suit and 52 the number of cards in a pack! In other words, we have boosted our STM storage capacity by making use of information already stored in LTM and by **elaboration** of its meaning. Similarly, it is much easier to remember items which have been **organised** into groups. If we were trying to remember cards which had been played during a card game we would find it much easier to remember them by suit (clubs, diamonds, hearts and spades) than randomly. These techniques of **chunking**, **elaboration** and **organisation** are not as silly as they seem. Memory experts make great use of such memory aids (mnemonics) to memorise long strings of numbers and names.

Storage

Our STM can only store information for a matter of seconds. If any extra information is added then items in STM are **displaced** (thrown out). Our STM is constantly being bombarded by new information as we talk, listen, watch or read, so material is constantly displaced from STM. If we want to retain the information for any length of time it must be passed to LTM. The displacement of information from STM can be illustrated by a simple party game. People are shown a series of around 20 familiar objects one after the other. As soon as they have seen all the objects, they are asked to write down as many as they can remember. It is found that the objects viewed both early and late in the sequence are best remembered.

The first objects are remembered better because the brain has had time to rehearse them before the STM is full. Those viewed late in the sequence are relatively easy to remember because they have not been displaced from STM. Objects viewed in the middle of the sequence have been displaced by others before the brain has had time to rehearse them. This is called the **serial position effect**.

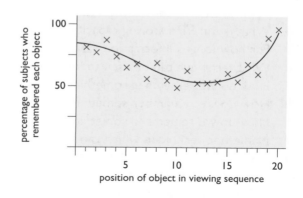

Graph to show serial position effect

Retrieval

Although we know we have remembered something, it is not always easy to retrieve it from our memory. How often have we said 'It's on the tip of my tongue'? Sometimes it is easier to retrieve a particular fact or episode if we find ourselves in the same situation, or context, as when we encoded it. **Contextual cues** which aid recall might be smells, sounds or the sight of a particular place or person. In these situations we sometimes say that 'our memories flooded back to us'. However, we are rarely in the position of being able to return to the place where the memory was formed! If we are having difficulty remembering something we have forgotten, then the recall of the item is often aided by thinking around the situation in which it was stored. We may be trying to remember the name of someone we met at a party. Often, if we visualise the evening out, the drive to the party, the house it was in and the others whom we met, the name may be located in our memory more easily. Sometimes retrieval is aided by changing the subject entirely and leaving the brain to search for the item subconsciously.

Questions

21 What is the principal function of the limbic system?
22 What kind of substance is NMDA?
23 Name a disorder common in old people in which severe memory loss is a symptom.
24 What are the three stages in memorising and recalling information from memory?
25 Approximately how many items of information can be retained in short term memory (STM)?
26 What is the STM storing capacity called?
27 For how long is information stored in STM?
28 What technique can be used to improve the memory capacity of STM?
29 How do we remember something if we do so by acoustic coding?
30 How do we remember something if we do so by visual coding?
31 How do we remember something if we do so by semantic coding?
32 Name one technique which can be used to help store information in LTM.
33 If a long sequential list of objects is memorised, which objects are most easily remembered?
34 What is this effect called?
35 What is a contextual cue?

THE DEVELOPMENT OF BEHAVIOUR

All aspects of development, including behaviour, are influenced by **maturation**, **inheritance** and the **environment**. The central problem of studying behaviour is the degree to which each of these factors influences behaviour.

Maturation

The development of many aspects of human behaviour is related to our age. Newborn children cannot speak, walk or grasp a ball but, as they mature, they develop the ability to do all these things. The sequence in which these developments occurs is fairly constant even between children of different races and cultures. The table opposite gives the age at which 50% of young children reach the various stages in their development.

STAGE OF DEVELOPMENT	AGE AT WHICH 50% REACH STAGE
sits without support	5+ months
stands with support	6 months
crawls	8 months
walks with support	9 months
walks independently	1 year

It is very difficult to cause deviation from these stages of development, even through fairly rigorous training, and so we can conclude that the behaviours have a genetic basis and are linked very closely to age. We know, for example, the nerve fibres of very young children are not fully **myelinated** and so the nerve impulses travel more slowly in the axons.

Inheritance

It is quite clear that inheritance plays a major part in our behaviour. For example, such conditions as phenylketonuria (PKU) and Huntington's chorea (HC) are controlled by single genes and both conditions result in a deterioration of the nervous system, with consequent changes in behaviour. PKU children who remain untreated suffer severe retardation in mental activity and die before the age of 30. HC victims, although apparently perfectly normal for the first 30 years or so of their lives, gradually lose their ability to talk and control their movements, and they show a marked deterioration in memory and mental ability before they eventually die.

There are many examples of 'normal' behaviour in humans too, which clearly have a genetic basis: territorial behaviour, sexual behaviour and aggression to name but three. But can the influence of the environment override the influence of the genes? In many cases it can.

Environment

The influence of the environment on behaviour can best be studied using **monozygotic** (identical) twins and **dizygotic** (fraternal) twins, especially if they have been reared apart. If the twins are monozygotic, the effect of genes can be discounted because the twins will have an identical genotype. Any differences in behaviour must then result from differences in their environment only.

In the personality disorder schizophrenia it is known that if one twin is schizophrenic then the chances are high that the other twin will exhibit some signs of similar mental disturbance. But whether the second twin develops the full symptoms will depend on the environmental influences he or she has experienced whilst growing up. It seems that genes predispose (make more likely) individuals to certain behavioural characteristics, but whether or not they develop them further depends on their environment, especially during early development.

Intelligence

People differ in their intellectual ability, but how much of this is due to genes (nature) and how much is due to environmental differences (nurture) has been the subject of much controversy and debate. All that one can say is that both genes and the environment influence intellectual ability. Intelligence is one aspect of intellectual ability which has been studied extensively for many years. **Intelligence Quotient (IQ) tests** used to be regarded as more important than other measures of intellect, but they do not test fine motor skills, artistic skills, social and leadership skills or the ability to memorise large quantities of factual information. They may not even have much to do with imagination and creativity. In fact, there is no general agreement as to what constitutes intelligence. But IQ tests have provided a fund of information to shed light on the 'nature' versus 'nurture' debate. For example, can IQ be improved by practice or even by a change in diet? There is evidence to suggest that both are possible. Do identical twins have the same IQ even when separated at birth and reared apart? No they don't but their IQs are more closely similar to one another than to other relatives.

Intelligence Quotient (IQ) questions

Some questions rely on your ability to sort out the meaning of words.

For example, which two words in brackets are connected in the same way as those underlined?

cruel – kind (angry, generous, unhappy, timid, selfish, shy)
mouse – mammal (thought, sugar, taste, emotion, water, fear)

Some questions rely on your ability to spot sequences.

For example, what are x and y in the following sequences of numbers?

1, 2, 4, 7, 11, x 3, 5, 4, 6, y, 7, 6, 8, 7

And:

If 2 and 4 with 3 = 5;   4 and 3 with 2 = 10, what does 2 and 3 with 4 equal?

Some are more subtle and rely on more devious powers of deduction.

For example, what is the next letter in each of the following sequences?

o, t, t, f, f, s, s, e, n, ...   m, t, w, t, f, s, s, ...   c, f, i, l, o, r, ...

(The answers are at the back of the book, on *page 99*.)

INFANT ATTACHMENT

Human children spend a much longer time 'growing up' than other young mammals do. The evolution of sophisticated communication skills is unparalleled in other species. This has resulted in the need for an extremely long period of time for learning and consequent **dependency**. From the moment they are born, children are highly dependent on their parents and this dependency requires strong bonds to be forged between the infant and its parents at a very early stage. Crying, smiling and suckling all provoke powerful responses in the parents. This attachment can also be seen in many other higher animals and has obvious survival advantages, as the young are more likely to obtain food, warmth, comfort and protection from their parents as they grow. But in humans the extra time required is for the enormous amount of learning which takes place.

One might guess that the most important of these early needs would be food, but experiments with monkeys often show that young monkeys rate warmth and body contact higher than food. Infant monkeys separated from their mothers at birth will choose a warm, cloth model which provides no food, in preference to a cold wire model which provides them with milk. If the model rocks, the young monkey will be even more attached to it! When all these features are tested separately, it appears that contact with a soft cuddly model is preferred by infant monkeys.

However, even a soft warm model mother which rocks and provides food is not enough to ensure infant monkeys develop satisfactory 'normal' behaviour. Young monkeys reared in such an artificial situation have difficulty coping with other monkeys and show inappropriate sexual behaviour. What is more, females raised in this way turn out to be poor mothers, tending to neglect, or even harm, their infants.

It would be dangerous to assume that everything described above applies to humans, but equally it would be foolish to assume that the general principles are not the same. It is certainly true to say that the links a child makes with its mother and, to a lesser extent, its father, in its early years, are vital in establishing its ability to cope with adult life, to establish sound relationships with others, and to be an effective parent.

There appears to be a relationship between the pattern of early attachment in humans and the way an infant copes with new experiences in subsequent years. Children who have a secure upbringing very often approach problems with enthusiasm and persistence. When they encounter difficulties they seldom give up and do not show many signs of frustration. Those who have experienced insecure attachment as infants become easily frustrated, give up readily and do not seek or welcome adult assistance. Children observed as securely attached early in life tend to be social leaders and are sought out by other children. Teachers rate them as self-sufficient and eager to learn. Insecure children tend to be socially less able and reluctant to participate in group activities. Teachers see them as less curious and less forceful in pursuing their goals, and these differences are not related to intelligence. These trends are general, of course, and there are many individual differences between children with otherwise similar backgrounds and experiences.

COMMUNICATION

Communication is the transfer of information from one individual to another. All animals can communicate, and some do so in sophisticated ways, but none can approach the amazing complexity of human communication skills. It seems certain that our **innate** ability to learn language is unique to our species. Many humans cannot read and many cannot carry out simple arithmetic, but almost all can easily communicate by use of complex spoken language.

Infants, before they are able to speak, communicate most effectively in a non-verbal way with their mothers. The infant smile is a particularly powerful signal to its mother. However, **non-verbal** communication takes place between adults too, often without us being aware of it. Subtle movements of the body and the face give much information to others about our mood, intentions or our deceptions, even when we try to convey the opposite information through language! In fact, when verbal signals contradict non-verbal signals, the effect can be most powerful.

Interestingly, some non-verbal signals vary in the information they convey from one culture to another. In Russia the slow hand clap is complimentary. In Italy the subtle difference between the two-fingered gesture (V-sign), palm out or in, is not recognised, so Italians usually respond positively to the V-sign whether given palm out or in! The thumbs up signal we use to attempt to gain a lift from a passing car is translated to 'sit on this' by Sardinian drivers!

All parts of our body are used in non-verbal communication – the way we stand, the use of our hands, even the clothes we wear signal what we are. In schools where uniform is worn it is still quite possible for students to convey information about themselves by the manner in which they wear their uniform.

However, when we communicate with others we invariably find their faces most interesting. Most of human non-verbal signalling is carried out using the multitude of facial muscles which we have acquired through millions of years of evolution. Even the whites of our eyes have evolved so that eye movements are detected more easily by others. The human face is far more mobile than that of any other primate. In particular, the eyes and the mouth convey a huge variety of non-verbal information. When we look at someone who is speaking to us, our eyes constantly check their eyes and mouth more than any other part of their face.

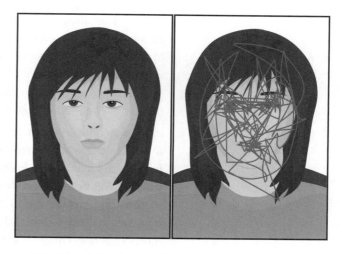

We do not look at only one part of a person's face when we look at their face for several minutes. Instead, we look around the person's face, spending different amounts of time on different parts, with most time spent on the eyes and mouth. This example shows what parts one person looked at over several minutes of looking at this face.

LANGUAGE

A great deal of our information about brain mechanisms for language comes from observations of patients suffering from brain damage. The damage may be due to tumours, penetrative head wounds, or the rupture of blood vessels. From these studies we have been able to find out which areas of the brain deal with the various aspects of speech and communication in general. For example, we know that the left side of our brain governs the ability to express ourselves in complex language, and the right hemisphere can only comprehend very simple language.

All languages have a limited number of speech sounds. In English we distinguish around 40 different sounds but an average adult will understand around 40 000 words. In addition, rules for combining words make it possible for humans to understand billions of sentences, and by the age of five we have accomplished most of the task! Children can speak a few isolated words at the age of 1; by the age of 2 they can speak in two- or three-word sentences; by the age of 3 the child's grammar is becoming more complex, and by the age of 4 the sentences are quite sophisticated. By the age of 6 most children have a vocabulary in excess of 15 000 words.

Can other primates communicate in the same way? Attempts have been made to try to get chimpanzees or gorillas to learn to speak. Speaking is beyond them because of their lack of fine motor control of voice box, tongues and lips. However, it is possible to teach them to communicate by sign language of one kind or another. There are cases of apes being taught a few hundred signs or signals, but it appears they cannot construct sentences with these symbols. Moreover, humans can place the words 'chimpanzee', 'chased', 'the' and 'man' into a sentence: '*The chimpanzee chased the man.*' whereas apes cannot. What is more, humans can switch the same words around to convey a different meaning: '*The man chased the chimpanzee.*'

How do children learn to speak? Do they simply imitate adults, or is there more to it than this? **Imitation** certainly plays a part in learning. That is why French children speak French and why English children speak English. Moreover, parents will point to an item such as a spoon and repeat the name of the item to the child so that it can attempt to imitate the sound. There is much evidence, however, to show that children learn to produce sentences by quite a different method. Young children are constantly producing meaningful sentences which they quite clearly have never heard before, e.g. '*All gone milk!*' In addition they generate words which do not exist: '*Daddy taked me to swim pool.*' This shows that they are applying general grammatical principles rather than copying what adults say. In reality, it is impossible for children to copy everything they hear. It is estimated that the number of 20-word sentences that we can understand is of the order of 10^{20}. To learn such sentences by repetition would take a child many billions of years if the sentences were simply spoken once each at normal rate! There is much other evidence (from comparison of different languages and from studies of children brought up in an environment where no one speaks) to show that the learning of a language is a complex process, much of which is innate. Moreover, the processes are quite unique to humans. This powerful tool which we have at our disposal has resulted in our rapid cultural and scientific evolution.

THE EFFECT OF EXPERIENCE

If a hungry cat is placed in a cage with a simple latch and a piece of fish is left outside, it may eventually learn to lift the latch to get to the fish. If this situation is repeated often enough, the cat will learn to respond very quickly. It might appear that the cat has solved the problem in the same way as a human might look around, assess the situation, and make intelligent guesses as to what to do. But cats are not as smart as humans and, even if the cat's paw is put on the latch by the experimenter a few times, the cat seems unable to make any kind of logical deduction. It is unable to **imitate** our action. Instead, the cat hits upon the solution by accident, and it is only after it has solved the problem in this way a number of times that it starts to associate its action with the reward. This kind of behaviour, performed by many animals, including ourselves sometimes, is called **trial and error** behaviour. The type of reward, the frequency with which the reward is given, and the time interval between response and reward, all have an effect on the behaviour of the cat – and us for that matter! Instant rewards **reinforce** the behaviour better than rewards which follow some time later.

Strangely, intermittent rewards reinforce activity more than regular rewards. Perhaps this is why fruit machines are so addictive! If rewards cease, then the behaviour undergoes **extinction**, i.e. it disappears.

If the cat learns that the lever will only open the door when a light comes on in its cage, then it has learned to **discriminate**. If the cat is taught to perform an increasingly complex series of tricks by an experimenter, this process is called **shaping**. Each action that the cat performs which approximates to the wishes of the trainer is rewarded, and progress is made towards the ultimate goal of the trainer.

Animals can also be taught to behave in particular ways by punishing them for the wrong action, rather than rewarding them for the right action. This form of reinforcement can be less effective, because the animal only learns what *not* to do, rather than what to do. But in some circumstances, pain can be a very powerful motivational force.

Much can be learned about animals' responses to rewards and punishment, and such knowledge can be related to human behaviour. Young children respond better to instant intermittent rewards for good behaviour rather than delayed punishments for bad behaviour. This is not an argument for abandoning punishment regimes however! It is not hard to imagine situations where a punishment rather than a reward fits the action. In truth, many reward children or animals at the wrong time and promote bad behaviour as a consequence. We must beware of giving in to a dog's or a child's demands with some form of reward, because we are producing quite the opposite behaviour from that which we intend! An example of this might be the mother who gives her child a sweet to stop her crying at the supermarket checkout. The child quickly learns that crying at this point in the shopping trip provides a tasty reward. If the child is ignored or rebuked at this point, then the behaviour will probably become extinct (for ever one hopes!). But many parents still take the soft option and everyone suffers – not to mention the damage caused to the child's teeth! Similarly, if a child cries in bed at night until

someone comes to comfort it, then this behaviour is reinforced. However, it is not wise to ignore crying children, particularly if they are very young. There may be a very good cause for a child to cry and hence grab our attention. What is more our genes have programmed us to be very conscious of such powerful signals, and not without good reason.

If a child is bitten by a dog, or scratched by a cat, it may cry and it would be wrong to ignore its cries. Following this incident, the child may then be frightened by all dogs or cats. If this happens we say that the child's behaviour has become **generalised**. The opposite type of behaviour is called **discrimination**, where a child might be taught that the family pet is quite safe to handle but the Rottweiler down the road is not!

Humans are particularly good at **imitation**. In fact, a great deal of human behaviour is learned by observing and imitating the behaviour of others. This is often a preferred means of learning and is used in training, e.g. when learning to ride a bike or play a piano. Here, **practice** plays a key role in establishing the **motor skills** required. By continued imitation and practice, neural pathways are established in the brain which can remain in place for the rest of one's life.

GROUP BEHAVIOUR

Humans are social animals with very complex cultural rules and a great deal of human behaviour results from influence by others. Our parents have the most profound influence on our behaviour, followed by brothers and sisters (siblings), friends of the same age (peers) and teachers. We are also influenced by people we admire (e.g. by film stars and sports stars) and we are influenced by what people do. If we were not, then advertisers would cease to advertise.

We may change our behaviour (e.g. stop smoking) because of something we have heard, seen or read, which convinces us that it is indeed worth changing, for good reasons which we believe to be true (better health in the case of smoking). This more permanent alteration of behaviour is known as **internalisation**.

We may change our behaviour, not because we believe in any intrinsic value in the behaviour we wish to copy but because we know someone whom we admire or respect behaves in this way and we wish to be like them. This alteration of behaviour is known as **identification**. A boy or girl might take up a sport played by a person he or she admires. Trends in fashion are often brought about by our wish to be like people we admire.

We may perform a task more effectively in the presence of others than we might do on our own. This alteration of behaviour is known as **social facilitation**. It is worth noting, however, that the complexity of the task influences the outcome of this form of behaviour. In general, people will run faster if watched by spectators. They will also tend to run faster if accompanied by another runner. However, although people will perform more arithmetical problems in a given time when others compete against them, they will produce more errors than they would if they had completed the activity on their own! We might simply say 'they were nervous!'

We behave differently when we are with a crowd of others. In such situations we can find our behaviour quite different from what we would consider as being normally acceptable. This alteration of behaviour is known as **deindividuation**. Deindividuation can result in unpleasant and disturbing human behaviour, for example, in the case of aggressive behaviour of football supporters. In these situations, the behaviour of others in the group is copied by individuals to ensure their acceptance by the group. Antisocial behaviour by the individual is thought to be admired by others and be less likely to be discovered or punished.

Questions

36 Aspects of development are influenced by maturity, environment and what other factor?

37 Why do nerve impulses not travel so fast in very young children?

38 Why are monozygotic twins useful for behavioural studies?

39 Why do children have a long period of dependency?

40 Give an example of a non-verbal signal.

41 What kind of learning takes place when we try various solutions and find the answer by accident?

42 What term is used to describe the reward of behaviour which is close to, but not exactly, the required behaviour?

43 What is the difference between generalisation and discrimination in terms of being frightened of snakes?

44 Distinguish between internalisation and identification in terms of trying to stop smoking cigarettes.

45 What term is used to describe improved performance in the presence of others?

46 What term is used to describe atypical behaviour in the presence of a crowd of other people?

THE POPULATION EXPLOSION

It is probably true to say that most of the major problems we face today are a direct result of the fact that there are increasing numbers of humans in most parts of the world.

The growth of the human population is not a new phenomenon. It has been going on since the emergence of our species less than a million years ago. For most of our evolutionary history, our population has shown slow but consistent growth. However the rate of growth has become explosive in the past 50 years and has only recently shown any sign of decline (see *page 83*).

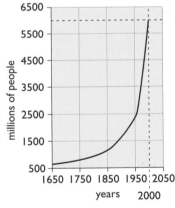

Because populations in some developed countries, such as our own, are relatively stable and because the greatest population increase is taking place in the poorer countries, it would be easy for us to say that the blame lies elsewhere. But it must be remembered that the population densities of some of the most advanced countries of the world are very high and their consumption of resources is much higher than that of the poorer countries. The United States has less than 5% of the world population but consumes 30% of the world's energy resources. India supports about 17% of the world's population but only uses 2% of its energy. Hong Kong has a population density of 6400 per km². Europe has a population density of 100 per km². But Africa has a population density of only 28 per km².

So what are the reasons for the increase in the human population? The first small populations of human ancestors appeared in Africa more than four million years ago. Since then, humans have spread out to inhabit practically every land area on Earth. Today there are over 6 600 000 000 people inhabiting the Earth and the population is increasing by about 1% per annum (an increase of about

70 million per year). This may not sound very much, but if this rate of increase of two persons per second continues for the next 700 years there will only be enough room for one person per metre^2 of the entire land surface of the Earth!

Before the invention of agriculture our ancestors depended on hunting and gathering as a means of obtaining food. Consequently, their population was regulated by what they were able to obtain from their environment and our estimates of their population size are based on our knowledge of the optimum numbers that can be sustained by this way of life, i.e. by knowing the **carrying capacity** of the Earth. By comparison with the populations of hunter-gatherer societies of humans living in the jungles or in the deserts today, it seems that around five to ten million humans inhabited the Earth before the agricultural revolution. This is around the same number of people who live in Scotland today! Primitive human populations, like all animal populations, were held in check by predation, availability of food, disease and climatic factors such as cold and drought.

Moreover, nature operated its own form of contraception. Young children were carried by their mothers for many years and could suckle on demand almost to the age of five. Suckling stimulates the production of hormones which inhibit ovulation, so women were unable to become pregnant for a number of years after the birth of each child. Today, many babies are bottle fed, and those who are breast fed cannot always obtain a drink when they wish, so this natural form of contraception has virtually disappeared from all human societies.

It is likely that the human population only increased very gradually for many thousands of years. However, about ten thousand years ago the population started to increase more dramatically as humans learned to cultivate crops and domesticate animals – the beginnings of **agriculture**. By about 2000 years ago, the world population had reached around 200 million, and it doubled in the next thousand years. Since then the rate of increase has been even more rapid and, at the present rate of increase, the world population is doubling not every thousand years but every forty years.

The **Agricultural Revolution**, which started ten thousand years ago, resulted in a huge increase in food production which continues today to outstrip population growth in most parts of the world. For this reason, many argue, we should not be concerned about population growth.

The **Industrial Revolution**, which started in the 1800s, brought with it great improvements in both agriculture and medicine: it improved **food supply** and improved **health care** and resulted in a massive decline in the death rate, particularly of young children. Unfortunately, then and now, birth rates in many parts of the world have changed little since humans first appeared on Earth.

It is important to realise that most animal and plant populations do not show long term growth of the kind we see in humans. Instead, animal and plant populations remain relatively stable for long periods of time, fluctuating only within narrow limits. Major changes in populations normally only take place in response to marked changes in the environment's capacity to support a particular species – its **carrying capacity**. Furthermore, most populations are stable, in the sense that if they are affected by some environmental event, a fire or a particularly cold winter for example, they quickly return to their previous size when conditions return to normal. It follows, therefore, that in such stable populations, **birth rate** matches **death rate**, or some **emigration** or **immigration** occurs to balance the books. In a growing population, such as our own, birth rate exceeds death rate. If we consider the human population of the world as a whole, then emigration and immigration have no effect on the total population unless we board spaceships and leave the Earth!

So what solutions are on offer? In Europe and North America, the human populations are relatively stable, and some are even in decline. There has been a **demographic** (relating to population) change from high birth rate and high death rate to low birth rate and low death rate. The lower death rate has been brought about by improvements in medicine and sanitation; the lower birth rate has resulted from the availability of increasingly effective forms of contraception and abortion – whether we approve or not. The decision to limit family size has been mainly due to simple economics. In primitive agricultural societies, children were regarded as an economic bonus (more farm workers) and, since many died of disease, it was important for a family to produce as many babies as possible. However, as societies became industrialised and medicines improved, children were more likely to survive and be an economic cost (rather than an economic bonus) to the family. Today in Western society, children could be regarded as very expensive luxuries! At present, the cost of raising children is, by and large, the only factor limiting population growth in developed countries. In underdeveloped countries population growth is still considerable, with a few exceptions. For example, in China and India, where the human populations are enormous, the governments have introduced a number of laws and financial incentives in an attempt to limit population growth. So the future looks uncertain, but one thing is absolutely certain, if the human species does not reduce its birth rate voluntarily, or by compulsion, then early death must follow for many.

FEEDING THE HUMAN POPULATION

One of the most important factors which has permitted the human population to grow rapidly, particularly over the last two hundred years, is the revolution in the production and transportation of food. Ten thousand years ago, humans started to tame wild animals and breed them for meat and milk and easy handling. In addition, they started to cultivate wild plants in 'fields' from which other plants (and animals) were excluded – as **monocultures**. Natural vegetation was removed to allow for this new-found activity – farming. As a consequence, there was no need for a nomadic life and humans could settle in communities – villages, towns, cities and now, 'mega-cities' containing millions of humans.

To feed the growing human population:

- forests continue to be cut down to make way for fields
- deserts are made fertile with irrigation schemes by building wells, dams and canals
- mountain sides are terraced for the growing of crops such as rice
- areas of land are placed under glass to protect them from weather extremes and to raise temperatures artificially
- the sea is kept at bay with dikes, and the salt washed from the mud to enable crops to be grown
- crop rotation is employed to reduce uptake of the same minerals from the soil from year to year and to keep one step ahead of invading pests
- artificial fertilisers (nitrates and phosphates) are used to raise the productivity of poor soil
- pesticides are constantly improved to stay one step ahead of mutating insects, fungi, bacteria and viruses which damage our crops and kill our animals
- herbicides are used to kill unwanted plants
- hormones are developed to increase growth rates
- antibiotics are used to kill animal pathogens
- large numbers of animals, such as chickens and pigs, are kept under one roof and reared intensively, often with scant regard for their well-being
- new machinery is invented to speed up planting and harvesting of crops
- refrigeration techniques are used in the transport and storage of produce

… there is no end to human ingenuity.

Selective breeding is one area where enormous advances have been made recently in the development of new strains of plants and animals. For thousands of years, humans have selected the best seed and the best young animals and bred them to produce desirable traits such as disease resistance, fast growth, good yield, flavour, colour and so on. However, recently the pace of artificial evolution has quickened with the development of **genetic engineering**. Genes can now be selected for valuable traits and transferred from one organism to another. Breeding can take place in the test tube or petri dish and **clones** (genetic copies) of animals and plants can be produced very quickly and relatively easily.

Despite all the advances, however, the provision of food for the world population is very uneven and many millions live at starvation level while others, in more technologically advanced societies, produce a superabundance of high quality food. In these countries, farmers and fishermen are paid *not* to produce more food, and vast stocks are stored in huge warehouses. The transport of surplus food would be relatively easy, but the costs are such that the wealthy nations, rightly or wrongly, are reluctant to be so generous. What is more, the provision of food on its own is regarded by some as unwise as it offers only temporary respite and simply encourages further population growth in areas already unable to sustain their current populations.

Tragically, one thing is certain: despite all the technological advances and the overproduction of food in certain areas of the world, many millions are yet to die of **malnutrition** and **starvation** before a global solution is achieved.

Genetic modification

Mention foods produced using genetic modification (GM) and most people express suspicion and hostility. Yet, some would argue there is nothing to worry about. So, what are the arguments for and against?

Farmers and scientists have crossed different varieties of animals and plants for many years, discarding the offspring they don't want and breeding from the ones that have suitable characteristics. As a consequence we have a huge variety of crop plants and domesticated animals to choose from.

However, in recent years we have been able to take genes from one species and transfer them to a different species to produce a desired effect. This is a huge advance because before we could only shuffle genes within one species and hope for the best. Moreover, now we can be very precise: we only move genes we want to move and we don't leave the whole breeding process to blind chance.

So why is there so much opposition to the process? Those with religious beliefs say we should not be tinkering with God's creation. Others argue that genetically engineered organisms may pass their altered genes to wild species and bring about unpredicted changes. Also, can we be certain that genetically engineered plants and animals are safe to eat? There is much to consider, but at the end of the day, the choice is yours to make.

Body mass index

A major health issue for modern society is obesity. A large proportion of people in Scotland are now overweight or obese, and the proportion is increasing every year. Obesity can lead to all kinds of health problems in later life, such as heart disease and diabetes. Some even predict that average life expectancy will decline as more people die early from conditions connected with obesity.

Various methods can be used to find out if you are overweight, and one, commonly used, is the body mass index (BMI). This a rough comparison of height and body weight. The calculation must be treated with caution, especially if you are a well-muscled male rugby player or weight lifter, because muscle is more dense than fat and raises BMI.

The formula for calculating your BMI is: mass (kg) / height (m)^2

Example: if you are 1·70 m high and weigh 60 kg your BMI would be 20·7 (5 ft 7 in and 9·4 stone)

The table shows the ranges of BMI and their description.

BMI	DESCRIPTION
less than 19	underweight
19–24	normal
25–29	overweight
30–39	obese
over 40	severely obese

WATER SUPPLY AND THE ENVIRONMENT

The worldwide need for a supply of fresh water is increasing for two simple reasons:

- there are more humans on the planet
- each human is using more water per day, directly or indirectly.

Not surprisingly perhaps, the developed countries are those with the greatest demands. In the poorest countries, the consumption per person is less than 10 litres per day, whereas in the UK the figure is nearer 500 litres per day. The reason for this difference is that the UK, like other developed countries, uses a large proportion of its fresh water supplies to cool power stations, to irrigate fields and to supply industry. Also, each household uses much more water per day. For example, in the past 30 years the household consumption of water per person in the UK has almost doubled as more water is used to wash people, clothes, dishes and cars, and water gardens.

As the world becomes increasingly industrialised, the global requirement for supplies of fresh water will continue to rise. However, there are two major problems which result from this increasing industrialisation:

1   As forests are removed, many areas, which once had an adequate rainfall, are now experiencing drought. Over 50% of the rain landing on a tropical forest evaporates back to the atmosphere to form clouds and later, rain. If the forest is removed, the water, instead of evaporating, runs down the hillsides to the sea. What is more, the foliage and roots of the trees protect and bind the soil. So, when the trees are removed, the soil is exposed to the drying effect of the sun and the full force of rain. The soil is blown away by the wind or washed away by the rain, leaving barren **marginal land** which is unable to support crops for more than a year or two. Dams and irrigation channels, built to preserve and redistribute water, then become clogged by the silt and are useless within a few decades of their construction.

2   As the Industrial Revolution spreads across the world, there is a consequent increase in the pollution of fresh water and sea water. Poisonous gases from power stations and car exhausts dissolve in rain making it both acidic and toxic. The rivers and oceans of the world are still regarded by most people as convenient dumping grounds for untreated sewage and industrial waste. Pesticides (such as DDT) and fertilisers are washed from agricultural land to pollute rivers and lochs. Poisons enter the **food chains** and can be passed along from animal to animal, becoming increasingly concentrated at each link in the chain. At the end of the food chain the poisons are so concentrated that the animals are severely affected and may become infertile or die. Many of our birds of prey have been affected this way and, more importantly, we humans too are at the end of many food chains.

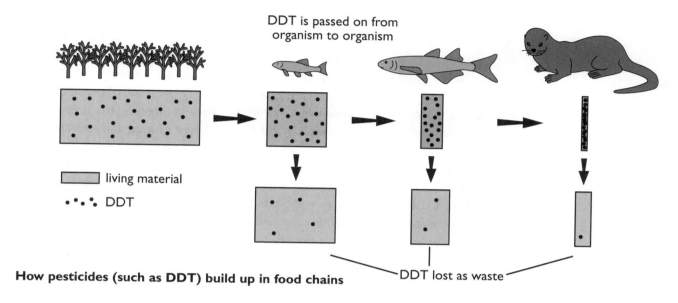

DDT is passed on from organism to organism

living material

DDT

How pesticides (such as DDT) build up in food chains

DDT lost as waste

It has been predicted that within the next 20 years water will have the same value as crude oil. Nations, in their desperation to secure dwindling supplies, will resort to conflict. For example, in the Middle East, Turkey controls the headwaters of both the Euphrates and the Tigris and has built over 30 dams. The Euphrates flows through Syria which plans to develop huge irrigation schemes for its agriculture. However, both rivers then flow through Iraq which also has a right to steady supplies. Similarly, the Nile flows through Uganda, Ethiopia and Sudan before reaching Egypt. Mexico City, with a population of 20 million, is sinking in places at a rate of 4cm per year as water is drawn from underground to supply the population. As a consequence, supply pipes are fracturing and vital water supplies are lost. Disastrous irrigation schemes to feed the Soviet cotton plantations have led to the Aral Sea in central Asia all but drying up. As it shrinks, poisonous salts in the water have become so concentrated that all the fish have died and the drinking-water is now toxic. Winds pick up the dried salts and dust and blow them over a wide area, contaminating crops and causing severe health problems to the population.

DISEASE AND THE HUMAN POPULATION

Without doubt, disease has limited the expansion of the human population much more than any other factor such as war or famine. The plague is a bacterial disease spread from rats to humans by fleas and from human to human by coughing. It is estimated that a third of the population of Europe was wiped out by an epidemic of the Black Death, as it was called, which started in 1347. For thousands of years, although disease killed a large proportion of the human population, this was compensated for by a high birth rate. Now, if birth rates remain high, but the treatment of disease improves, the potential for population increase is enormous. Certainly, the huge expansion in the human population over the past 200 years has resulted from our ability to control and treat disease effectively. This has been as a result of better **sanitation**, **hygiene**, **nutrition** and **health care**.

Two hundred years ago the **mortality rate** for infants in London was around 30% (300 per 1000). Today the rate is nearer 0·7% (7 per 1000). Then, life in the city was extremely unpleasant. Houses did not have bathrooms; the unpaved streets smelt disgusting as people walked in a mixture of mud and excrement; drinking-water from the Thames was heavily contaminated with sewage and people regarded it as unhealthy to wash. This was not surprising, because it was through water-borne diseases, such as cholera, dysentery and typhoid, that most died. Cures for diseases were often more dangerous than the disease itself. For example, venereal diseases were treated with mercury compounds. Mercury poisons the nervous system causing mental disorders and violent twitching. Surgery was extremely primitive. There were no anaesthetics and no knowledge of infection and how it occurred. If you were seriously injured you rarely recovered. Patients were knocked unconscious or given large quantities of alcohol before operations. Diet was very poor. Food became quickly contaminated as there was no refrigeration. Parasitic contamination of meat was common and nearly everyone was infected by intestinal worms. Tapeworms cause internal bleeding and can sometimes block the entire intestinal system. Food was usually washed down with alcohol and many suffered from kidney stones and gout as a consequence. Not surprisingly, life expectancy was very short.

However, in the 19th century there was a great advance in the understanding and treatment of disease. Smallpox, a viral disease, killed over a million people each year in Europe around the beginning of the 19th century. At that time, a British doctor, Edward Jenner, heard a milkmaid claim she could not catch smallpox because she had suffered cowpox. He then tested her hypothesis on a young child whom he infected first with cowpox, a mild disease, and then, at a later date, tried to infect with smallpox. The child remained healthy. The concept of **vaccination** ('vacca' is Latin for 'cow') quickly gained acceptance, and today smallpox has been completely eradicated by a worldwide programme of vaccination organised by the **World Health Organisation (WHO)**. Many other diseases, such as polio, measles, diphtheria, whooping cough and tuberculosis, are now rare in developed countries as a result of vaccination programmes.

Later in the 19th century the French chemist Louis Pasteur discovered that the souring of wine and milk was brought about by microbes and went on to suggest that microbes were the cause of disease rather than being the result of disease. He also developed vaccines for two deadly diseases, rabies and anthrax. Lister, an English surgeon, reading of Pasteur's discoveries, was the first to carry out operations in **antiseptic** (sterile) surroundings, towards the end of the century. At the same time Koch, a German scientist, identified many **pathogenic** (disease-causing) bacteria under microscopes and developed techniques for growing them on agar jelly.

Despite these great advances, many millions of people in underdeveloped countries still die of diseases which have, by and large, disappeared from developed countries. Water-borne diseases, such as cholera, dysentery and typhoid, are a particular problem resulting from poor sanitation and contamination of drinking water. Only through improved sewage disposal, the provision of clean drinking water, improved diet and widespread vaccination will these diseases eventually be brought under control.

The diagrams show two population pyramids, one for a developed country and the other for an underdeveloped country. In the developed countries (e.g. UK), **birth rates** and **death rates** are similar and infant mortality is low, so almost all the children born go on to survive to old age. In the underdeveloped countries (e.g. Bangladesh), many die young and the pyramid takes on a much more angular shape. The study of population statistics and the distribution of age groups within populations is called **demography**.

In the study of demography, birth rates and death rates play an important part. The bar graph below shows that there has been a dramatic decline in birth rates in countries all over the world over the last thirty years. The reasons for these changes are many, but they result principally from improvements in education, wider availability of contraceptives, changes to economic status and the continuing emancipation of women across the world. Just what impact this decline will have on the world population over the next hundred years, remains to be seen.

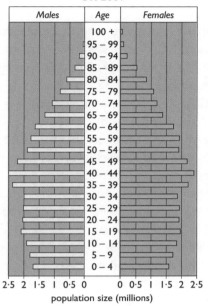

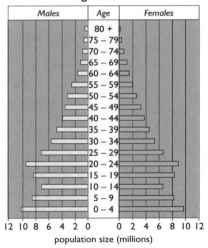

Population pyramids

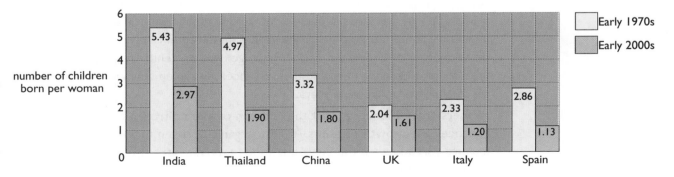

Decline in birthrate

Questions

47 What is meant by the carrying capacity of an environment?

48 What four factors will be in balance in a stable population?

49 What is the study of population statistics and trends called?

50 A population of 5000 is increasing at 2% per annum. What will be the population after 2 years?

51 Give an example of a monoculture.

52 What is a herbicide?

53 Why is the technique of crop rotation a worthwhile farming practice?

54 In terms of species involved, what is the advantage of genetic engineering over selective breeding?

55 Distinguish between malnutrition and starvation.

56 Why is there an increasing need for fresh water on this planet? Give two reasons.

57 What is meant by the term 'marginal land'?

58 What kind of substance is DDT and why is it particularly harmful to the environment?

59 Give two examples of improvements in health care over the past 100 years.

60 What is a pathogen?

THE NITROGEN CYCLE

Nitrogen is an essential element present in all nucleotides and amino acids. Without it, life as we know it would not exist. Almost 80% of the atmosphere is nitrogen – around 10^{15} tons of it – but, perhaps surprisingly, only a few organisms can make use of this massive supply of nitrogen directly. These organisms are mostly **bacteria** living freely in the soil or inside the roots of some plants, e.g. clover. The bacteria take in nitrogen from the air and convert it to nitrates and ammonia which can then be absorbed by plants through their roots. Animals obtain this combined nitrogen by eating either plants, or animals which have eaten plants. The nitrogen removed from the soil, however, is soon returned. Urine and faeces, and the bodies of dead plants and animals, are all degraded in the soil by other bacteria and fungi, which decompose the proteins and nucleic acids present in the organic material. The nitrogen in these compounds is then converted back to nitrates which can be absorbed by the plants again. In this way, nitrogen is recycled in an economical and efficient way by nature. Some nitrogen is lost inevitably, dissolved in rainwater and carried off to the sea or converted back to nitrogen gas by denitrifying bacteria (also found free living in the soil). In essence, the system has worked this way successfully for millions of years.

Farmers soon discovered that the addition of extra nitrogen to the soil enhanced the growth of plants. At first, organic nitrogen in the form of animal manure was added to the fields but soon inorganic nitrogen in the form of artificial fertiliser was manufactured using large quantities of energy from fossil fuels. It was inevitable that such a massive addition of nitrogen to the soils of

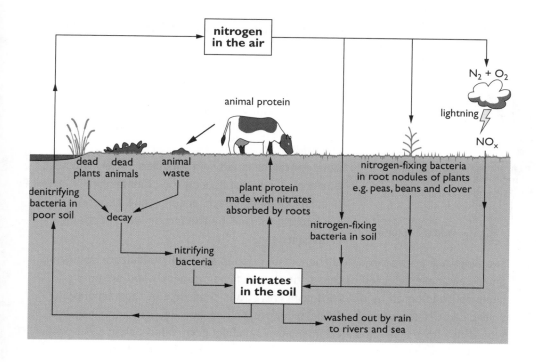

The nitrogen cycle

the world would disturb **ecosystems** and poison waterways and even the seas. Fertilisers and animal waste from intensive farming eventually find their way into rivers, lochs and the sea. There they have the same effect as they do on the land – they stimulate the growth of plants. Algae grow particularly well and the **algal blooms**, as they are called, have the effect, perhaps surprisingly, of reducing the oxygen content of the water in which they grow. Moreover, some algae are themselves toxic. The increase in organic material in the water is so great that many bacteria are able to feed off the dead remains of this huge **biomass**. In so doing they remove oxygen from the water to such an extent that other animals die, particularly at night, when photosynthesis does not contribute any oxygen to the ecosystem. This is known as eutrophication, and is now widespread throughout the world, affecting our lives in many ways. Fish stocks are depleted, swimming becomes dangerous for humans, and the excess nitrogen causes havoc to fragile ecosystems. For example, the Great Barrier Reef is threatened with destruction by starfish which thrive on the coral-forming polyps. It is thought that the huge increase in the number of starfish has been brought about by nutrient pollution from the mainland of Australia. Ammonia in the air, dissolved in rainwater, is now enriching barren moorland and affecting the growth of plants adapted over millions of years to a low nutrient environment. Of all these though, the factor which is most likely to encourage remedial action by humans, is the suspicion that dissolved nitrates in drinking water can have harmful effects on growing children. What these effects are is not certain yet, but there is some evidence that excess nitrogen may be carcinogenic (cancer-causing) and may also cause growth deformities.

An investigation into the effect of dissolved nitrates on plant growth

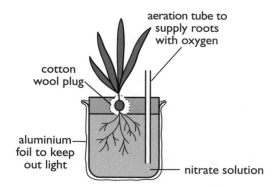

aeration tube to supply roots with oxygen

cotton wool plug

aluminium foil to keep out light

nitrate solution

It is possible to find out how nitrogen affects plant growth by growing plants in culture solutions. A series of containers is set up, like the one shown opposite, each with a different concentration of nitrate solution.

One container is left with distilled water only, as a **control**. A control is used for comparison, to prove that any changes which occur in the experiment are actually caused by the factor which is being investigated.

Seedlings are germinated and then suspended in each of the culture solutions. Small seeds are selected because they store fewer minerals and are more likely to be affected by different concentrations of nitrate in the surrounding water. Light is excluded from the solutions to prevent algal growth. The algae would use up the nitrates and upset the carefully controlled balance. All other variables are kept constant, such as temperature, pH, volume of solution, concentration of other dissolved minerals in solution, duration of experiment and type of seed. This is very important because if two factors are changed then it is impossible to tell which factor has affected the growth of the plants, and the results of the investigation would therefore be **invalid**. To ensure the results of the experiment are **reliable** a number of **replicates** (repeats) should be set up and averages calculated.

Growth of seedlings can be measured in a variety of ways: e.g. length of leaves, number of leaves, length of roots, dry mass of plant. The change in dry mass of plant material is a very good way of measuring growth, but unfortunately the plants have to be killed in this procedure, so their continued growth cannot be monitored.

THE CARBON CYCLE

All living things consist mainly of the elements carbon, hydrogen, oxygen and nitrogen, present in virtually the same proportions in all organisms from bacteria to humans (see table below).

ELEMENT	PERCENTAGE COMPOSITION BY MASS (EXCLUDING WATER)	
	bacterium	human
carbon	51	54
oxygen	20	19
hydrogen	10	8
nitrogen	10	9

Carbon is as essential for life as water. Because of its ability to share up to four electrons (valency) with other elements, it can combine with them in an almost limitless way. So carbon atoms are found in almost all the compounds made by living things, and such compounds are said to be 'organic'.

Carbon atoms get into the bodies of living things through the process of photosynthesis. All green plants remove carbon dioxide (CO_2) from the air, and water (H_2O) from the soil to manufacture glucose ($C_6H_{12}O_6$) using light energy obtained from the sun. The glucose is then converted, by the addition of various minerals such as nitrogen, phosphorous, potassium and sulphur, to the great variety of organic compounds needed by the plant.

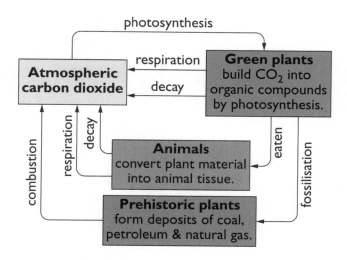

The carbon cycle

Animals then eat plants directly (herbivores) or indirectly (carnivores) to obtain their own supply of carbon compounds. In so doing they obtain much chemical energy. This energy is released during the process of respiration where the carbon is returned to the air as carbon dioxide. So carbon is continually recycled in a natural way through **photosynthesis** and **respiration**. Bacteria and fungi in the soil digest the dead organic matter and return the carbon to the air through **respiration**.

However, humans have disturbed this natural cycle by burning carbon compounds and releasing extra carbon dioxide into the air. Plants and animals living millions of years ago in swamps and peat lands where there was little decay have, over the centuries, turned to coal, oil and gas. These **fossil fuels** are rich in carbon compounds, e.g. **methane** (CH_4) which we burn to drive our industrial processes and warm our houses. Also, by burning the forests of the world we convert solid carbon, in the form of wood, to gaseous carbon in the form of carbon dioxide. Not surprisingly, this activity appears to be having an impact on the Earth's climate. The carbon dioxide, and other gases such as methane appear to be having a **greenhouse effect** on the planet. Methane is produced as a result of the metabolism of bacteria in the soil and in the guts of animals. The heat of the sun is not allowed to escape from the Earth quite so easily as before and **global warming** is occurring. This phenomenon is almost certainly the cause of the weather extremes we are now experiencing round the world – abnormal temperatures, storms, floods and droughts. If we continue to add greenhouse gases to the atmosphere it seems very likely that there are going to be even greater changes to the climates of the world; the increase in temperatures will cause icecaps to melt and seawater to expand and rise to flood areas where at present millions of humans are living.

Carbon emissions in 2005

COUNTRY	MILLION METRIC TONNES OF CARBON	METRIC TONNES OF CARBON (PER HEAD OF POPULATION)
USA	6200	22
UK	580	10
Scotland	52	10
Japan	1300	9
China	5800	5
India	1700	2
Zimbabwe	13	2

Data is changing rapidly from year to year as developing countries become increasingly industrialised.

Tropical forests

Many claim that the tropical forests are the 'lungs of the world' because they provide us with the oxygen we breathe. But this is not quite so. A mature forest is at equilibrium with its environment. It neither gains nor loses mass, so its output of oxygen is balanced by its requirement for oxygen. At night, all the organisms — bacteria, fungi, plants and animals — take in oxygen. Also, for 24 hours a day, there is a huge amount of decay, which matches the growth of new plants.

It is only when a young forest is growing and adding to its biomass that it removes carbon dioxide from the air and adds oxygen. Then, when it is mature, it represents a huge store of carbon. When we remove these forests by burning we release this stored carbon into the air as carbon dioxide, and, incidentally, remove a large volume of oxygen at the same time.

However, in the warm, damp jungles of the world, carbon dioxide for photosynthesis is often in short supply. So, as we add carbon dioxide to the air, plants are able to use it up more quickly and hence grow more quickly. Life is never simple.

Questions

61 What change to the environment is likely to result in an algal bloom?

62 Why is nitrogen important for animals and plants?

63 How do plants obtain their nitrogen?

64 How do animals obtain their nitrogen?

65 What kind of organisms trap atmospheric nitrogen and convert it to nitrate?

66 Name a species of plant which has root nodules.

67 What do denitrifying bacteria do with nitrates?

68 Why is there a loss of oxygen in water polluted by fertiliser run-off?

69 What is the function of a control in experimental situations?

70 What is the function of replicates in experimentation?

71 How many factors can be allowed to vary in an investigation if it is to be valid?

72 What major process, carried out by green plants, removes carbon dioxide from the air?

73 Why has there been a general increase in carbon dioxide concentrations in the atmosphere over the past 100 years?

74 Why is carbon dioxide referred to as a 'greenhouse' gas?

75 What effect is this having on the Earth's climate?

76 Name another greenhouse gas which results from bacterial activity.

77 Why are sea levels predicted to rise as a result of global warming?

The future of the planet?

As the number of humans inhabiting the world increases by the second, the destruction of the natural habitats affects the air we breathe, the food we eat, the water we drink and the climate in which we live. The precious soil is washed away and many species of animals and plants which might have been of great medicinal or food value are driven to extinction. Never before in the history of the Earth have so many wonderfully varied and beautiful ecosystems been damaged or destroyed so quickly or so effectively. The most destructive animal ever to have lived on the planet is conducting a giant global experiment which it is reluctant to halt and which risks its very existence.

TIPS FOR THE EXAM

> Use the following examples to help you prepare for the exam. The examples 'walk you through' some exam-style questions, giving you tips and showing how to answer certain types of questions.

UNIT 1: CELL FUNCTION AND INHERITANCE

Example 1

You must be flexible when it comes to dealing with diagrams. The examiners are always going to try to give a new angle to a topic, to test your total understanding.

The three-dimensional diagram of the nucleus and endoplasmic reticulum from the 2002 exam is a good example. It's likely you have never seen a diagram like this before, but if you have a good understanding about cell structure, this should not present you with any difficulties.

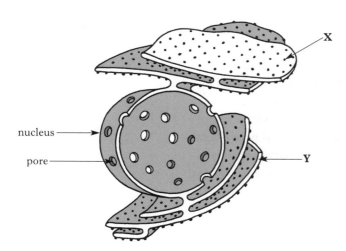

Sometimes you may find yourself in doubt about a labelled point. In the diagram, **X** could be pointing to a ribosome or the endoplasmic reticulum. How can you decide which? The simple answer is to say that **X** points to ribosomes on the endoplasmic reticulum; then you will definitely be given the mark.

However, the examiners have realised this could be a problem and have helped you in two ways.

- Question (a)(ii) asks: '*What kind of substance is manufactured by organelle **X**?*'
- Question (b)(i) states: '*The structure labelled **Y** is composed of sheets of membranes.*'

Both these confirm that **X** is a ribosome and that **Y** is the endoplasmic reticulum.

Example 2

You must be able to use your knowledge to solve problems in the exam. After all, that's what science is all about. The two questions below illustrate this.

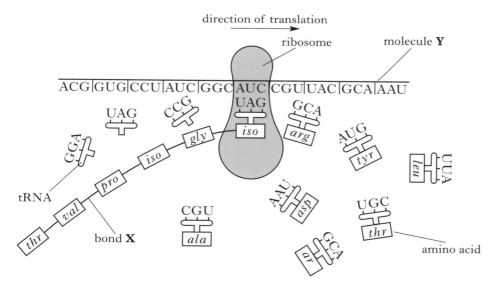

- Question (d) asks: '*What sequence of bases on a DNA molecule will code for the amino acid called thr?*' (Remember that *Thr* is the shorthand for Threonine.)

You can see from the diagram that the anti-codon for *thr* is UGC, which means that the mRNA codon is ACG. This is the answer that many candidates gave in the exam, but of course this code must then be transcribed back to the DNA code. So the answer is in fact TGC.

- Question (e) asks: '*Amino acids are added to the peptide chain at the rate of 15 per second. How long it will take for the complete synthesis of the peptide shown in the diagram above?*'

First, you must know that each amino acid is coded for by a triplet code. So the mRNA strand shown will code for ten amino acids. However, the rate is 15 per second, so the answer is 10÷15 of a second = **0·67 of a second**.

It's worthwhile remembering that the examiners are always very flexible and will accept 2/3 of a second, and even 0·6s or 0·7s, which is really quite generous.

UNIT 2: THE CONTINUATION OF LIFE

Example 1

Often an examination question will ask for a description of something. Always write a sentence or more when asked for a description. Also, check the number of marks available and match this with the number of facts you give. An example is given overleaf. There are three marks for this question, so you must give at least three points in your answer. Often it is a good idea to try to give more than three points, so that you can be sure of gaining full marks. The examiners have a list of points, all of which are acceptable.

Example

The hormone ADH is produced by the pituitary gland. Describe the role of ADH in restoring water balance after excess water has been lost from the body.

A good answer would be as follows:

The hypothalamus detects the water concentration of the blood. (1) When the body is short of water the pituitary gland secretes more ADH. (1) ADH acts on the collecting ducts of the nephrons (1) and makes them more permeable (1) so that more water is removed from the urine as it travels towards the ureter. (1)

Any three of these points would give you full marks for this question.

Example 2

The exam will always include graphs and ask you to extract information from them. These are problem-solving questions and they can be very testing.

An example is given below, from part of a question in the 2004 exam, in which a group of students take part in an investigation of blood pressure and pulse rate.

The graphs below show initial and final blood pressures of one of the students.

(i) Calculate the increase in the pulse rate of this student over the period of the investigation. You must use both graphs to answer this question.

Graph 1 Initial Blood Pressure **Graph 2 Final Blood Pressure**

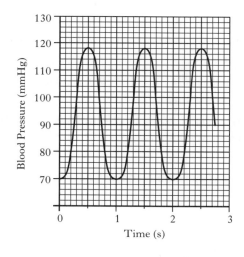 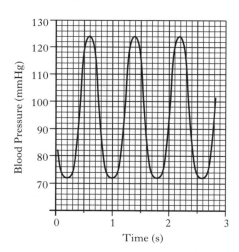

You can see from Graph I that the lines move up and down. This is the pulse. So these waves give you the pulse rate. Take any point on the graph and check the time difference between that point and another identical point. You will see that one heart beat lasts for I second. So the pulse rate is 60 bpm because there are 60 seconds in one minute.

From Graph 2 you can see that one heart beat lasts for 0·8 second. To calculate heart rate, always divide 60 by the number you are given: $60 \div 0.8 = 75$ bpm. So the heart rate has changed from 60 to 75bpm, an increase of 15bpm.

If you are given the heart rate in beats per minute and are asked the length of one heart beat, then carry out exactly the same calculation. Divide 60 by the number you are given and this will give you the answer. For example, if a person's heart is beating at 100 bpm, the length of one heart beat is: $60 \div 100 = 0.6$ second.

UNIT 3: BEHAVIOUR, POPULATION AND THE ENVIRONMENT

Example 1

The examiners will often start the page with an easy question, and then give you a more difficult question to separate the average candidates from the good candidates. Remember that around 25% of the examination includes problem-solving questions. The question below, taken from the 2003 exam, illustrates these points well.

The graph shows increases in brain volume at four stages of human evolution over the last four million years.

The bars indicate the range of volumes and the mid (median) value.

(a) State the range of brain volume for Homo habilis.

(b) Complete the table below for Homo sapiens.

SPECIES	MEDIAN VOLUME (CM³)	PERCENTAGE INCREASE
Australopethicus	500	-
Homo habilis	600	20%
Home erectus	800	33%
Homo sapiens		

The answer to (a) is 400–800 cm³. You must supply the units. If you don't, you lose a mark. The reading from the graph is not too easy because of the scale on the *y*-axis which goes up at 50 cm³ per box.

Question (b) is much more difficult because you need to be able to calculate percentages in addition to working out what figures to use for the calculation. Two answers are given to you to enable you to work out how this was done. So, you have to look at the answers for *H. habilis* and *H. erectus* to see how the calculation was carried out.

The 20% was calculated by taking 500 from 600 and dividing the answer by 500. This is how percentages are normally calculated, and you will always find percentage calculations in the exam:

$$\frac{\text{Change}}{\text{Starting value}} \times \frac{100}{1}$$

You can confirm this is the method by checking the second percentage: **200/600 x 100/1 = 33%**

You can now complete the last box by reading the median value from the graph for *H. sapiens*. This is 1400. So the percentage is the difference divided by the starting value, multiplied by 100: **600/800 x 100/1 = 75%**

Example 2

The following graphs appeared in the 2007 examination.

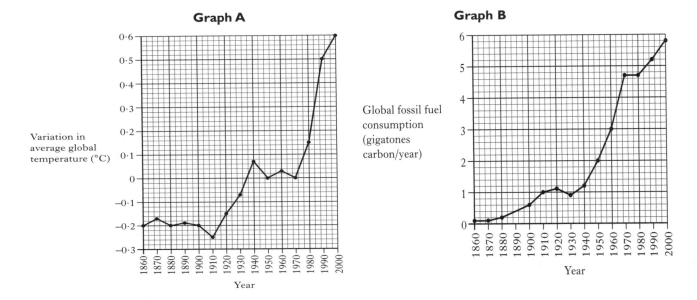

Graph A shows how the average global temperature, between 1860 and 2000, varied from that in 1970.

Graph B shows the global fossil fuel consumption, between 1860 and 2000.

One question asked: '*Discuss the extent to which the graphs support the theory that rising global temperatures are due to an increase in fossil fuels. Quote data from the graphs in your answer.*'

Very few candidates scored the 2 marks which were on offer. All spotted that the trends were similar, i.e. that as fossil fuel consumption has risen, so have average global temperatures. But that wasn't enough to gain even 1 mark.

To gain a mark, you had to spot that there were exceptions to the rule, e.g. between the years 1890 and 1910 average global temperatures declined, as they did between the years 1940 and 1950. These are against the trend.

Also, many candidates quoted data incorrectly. They forgot to include units, and they forgot to quote data from both axes, i.e. temperature variations *and* the years in which they occurred. They also quoted temperatures as if they were actual temperatures, whereas they were *variations* in temperatures.

So, be careful when asked to discuss data in the examination. You are meant to spot that all is not as straightforward as it first appears.

ANSWERS TO QUICK QUESTIONS

UNIT 1: CELL FUNCTION AND INHERITANCE

1   Metabolism
2   2mm (50 x 400 ÷ 1000)
3   On the ribosomes
4   Mitochondrion
5   Powerful digestive enzymes
6   They act as catalysts
7   Denatured
8   On the active site
9   They act as activators or co-enzymes
10 Competitive and non-competitive
11 Because, if they were active, they would digest the cell which produced them
12 The dark pigment of brown/black skin and hair.
13 Because an enzyme is missing in the metabolic pathway
14 Post-natal screening
15 Carbon, oxygen, hydrogen and nitrogen
16 Amino acids
17 Peptide bonds
18 Hydrogen bonds
19 ATP and DNA
20 They slide closer together
21 Deoxyribose
22 Nucleotides
23 Thymine
24 Guanine
25 Uracil
26 A triplet of bases on an mRNA molecule
27 12
28 To carry the base code from the DNA to the ribosomes
29 To bring amino acids to the ribosomes for attachment to one another
30 UGA
31 Peptide
32 To produce usable energy for the body
33 ATP
34 Glucose
35 Fats or amino acids
36 Up to 38
37 In the cytoplasm
38 Pyruvic acid
39 Krebs cycle
40 Carbon dioxide
41 It is a hydrogen carrier
42 On the cristae of the mitochondria
43 Oxygen
44 Water
45 Lactic acid
46 Glucose
47 Maltose
48 Because they are insoluble and have no osmotic effect
49 In the liver and muscles
50 Lipids
51 They have approximately two times more energy weight for weight
52 Glycerol and fatty acid
53 Oestrogen, progesterone and testosterone
54 Protein
55 Phospholipids and proteins
56 Fluid mosaic
57 Examples include: to act as self markers, to transport molecules from one side to the other, to act as hormone receptors
58 Osmosis
59 They burst
60 Active transport requires energy from the cell, and the molecules move up the concentration gradient
61 They both involve the infolding of the cell membrane
62 A substance (often a protein) which triggers an antibody response
63 Macrophage
64 Lysosomes
65 B-lymphocytes
66 Humoral response
67 T-lymphocytes
68 It is quicker and much more antibody is produced
69 Autoimmunity
70 In active immunity the body makes its own antibodies; in passive immunity the antibodies are provided ready-made
71 Through breast feeding, or through the placenta, mother to child

72 By being given a vaccination of a weakened germ

73 Nucleic acids and amino acids

74 Body cells (all cells except for gametes)

75 Karyogram

76 Length of chromosome and position of centromere

77 Homologous chromosomes

78 DNA replication > mitosis > cell division

79 Enzymes, ATP and a supply of nucleotides

80 Because half of the original strand is present in each of the new strands

81 Chromatids

82 Centriole

83 Meiosis: haploid cells which are not genetically identical; mitosis: genetically identical diploid cells

84 At the second division stage

85 46 and 23

86 So that when they fuse the original diploid number is regained

87 Chiasmata

88 So that offspring are varied (this aids survival in the long term)

89 Phenotype

90 50%

91 Co-dominant

92 A and/or B

93 The man has not inherited the allele, but the woman is a carrier. So, there is a 25% chance of them having a colour-blind child, and the child would be a boy

94 Polygenic inheritance

95 Non-disjunction

96 Substitution

97 One in which a piece of chromosome is removed, rotated and replaced

98 An organism which causes disease

99 Genome

100 To check for genetic abnormalities in embryos before birth

UNIT 2: THE CONTINUATION OF LIFE

1 Puberty

2 9 months

3 Pituitary gland

4 It stimulates a ball of cells, the Graafian follicle, to develop round the ovum, and it stimulates the follicle to produce oestrogen

5 Negative feedback

6 LH

7 At the middle – around 14 days after the beginning of menstruation

8 It becomes the corpus luteum

9 Progesterone

10 Oestrogen and progesterone

11 They inhibit it

12 It degenerates

13 It causes it to disintegrate, with consequent menstruation

14 In the seminiferous tubules of the testes

15 Testosterone

16 A steroid

17 Testosterone and FSH

18 It stimulates the production of testosterone

19 Prostate and seminal vesicles (also Cowper's gland)

20 She may receive hormone treatment

21 *In vitro* fertilisation

22 Oestrogen and progesterone

23 The vaginal mucus becomes thinner and the temperature rises by half a degree

24 A fertilised egg

25 Two sperm fertilise two eggs

26 Cleavage divisions

27 The lining of the uterus

28 In the oviduct (fallopian tube)

29 Around a week after fertilisation

30 Amniotic fluid

31 The mother may produce antibodies against the child's red blood cells

32 By diffusion

33 By active transport

34 Antibodies are carried by pinocytosis

35   Oestrogen and progesterone
36   Prolactin
37   It makes them contract
38   Colostrum
39   It stimulates the production of prolactin and oxytocin
40   1 : 3
41   Thyroxin
42   The SA :Vol ratio decreases
43   54 : 27 = 2 : 1
44   Artery
45   Artery
46   Vein
47   Capillary
48   Artery
49   Vein
50   (i)   Carotid artery
     (ii)   Pulmonary vein
     (iii)   Hepatic portal vein
     (iv)   Renal vein
51   120 bpm
52   75 bpm
53   Coronary artery
54   The left-hand side
55   Tricuspid valve
56   In the wall of the right atrium
57   Sino-atrial node
58   The autonomic nervous system
59   It speeds up heart rate
60   Medulla oblongata
61   Systole and diastole
62   One is systolic pressure and the other diastolic pressure
63   Vena cava
64   A thickening and hardening of the walls (atherosclerosis)
65   The fluid portion of the blood
66   Tissue fluid
67   The lymphatic system
68   It is cleaned – foreign particles are removed
69   Lymphocytes (and macrophages)
70   Plasma has much more protein in it
71   Nucleus and mitochondria
72   Haemoglobin

73   It enables them to move through narrow capillaries/it gives them a high SA :Vol ratio for easy transfer of oxygen in and out of the cells
74   120 days
75   Vitamin B_{12}
76   In the liver
77   It is converted to bilirubin
78   It increases its ability to release oxygen
79   An oxygen dissociation curve
80   20%
81   Breathing rate is increased
82   To increase the surface area of the small intestine for the absorption of food
83   Lacteal
84   Fats
85   Intrinsic factor
86   Deamination
87   Urea
88   In the gall bladder
89   Bile emulsifies (breaks up) fats
90   A, B_{12} and D
91   Nephrons
92   Ureter
93   Bowman's capsule
94   Proximal convoluted tubule
95   Glucose or amino acids
96   ADH
97   By the pituitary in response to a lack of water/ thirst
98   The maintenance of the steady state
99   Glucagon and insulin
100 Glucagon
101 Glycogen
102 Hypothalamus
103 They vasodilate
104 It must evaporate
105 The very young and the very old

UNIT 3: BEHAVIOUR, POPULATION AND THE ENVIRONMENT

1 A nerve cell
2 Cerebrum
3 Medulla oblongata
4 Muscle co-ordination and balance
5 Corpus callosum
6 Away from the nerve-cell body
7 Myelin sheath
8 It speeds up transmission
9 Medulla oblongata
10 Cerebrum
11 The brain and the spinal cord
12 The somatic system
13 The sympathetic system
14 Sensory, relay and motor
15 Diverging pathway
16 Rods are found on the retina and they are sensitive to light
17 A point where two nerve fibres meet
18 They are carried by a chemical called a neurotransmitter
19 Acetylcholine or noradrenalin
20 They are reabsorbed or broken down by enzymes
21 As a place to store memory
22 It is a receptor
23 Alzheimer's disease
24 Encoding > storage > retrieval
25 Seven
26 Memory span
27 For around 30 seconds
28 Chunking
29 By its sound
30 By picture/appearance
31 By meaning
32 By elaboration or organisation
33 The ones at the beginning and the end
34 Serial position effect
35 A stimulus which helps us to remember a related event

36 Inheritance
37 Because the nerve fibres are not fully myelinated
38 Because the effect of inheritance is identical in each
39 To give time to learn
40 Examples include smiling or frowning, waving a hand
41 Trial and error
42 Shaping
43 Generalisation: frightened of all snakes; discrimination: only frightened of poisonous snakes
44 Internalisation: stopping because we believe it makes sense; identification: stopping because we have seen someone we admire do the same
45 Social facilitation
46 Deindividuation
47 The population which an environment can sustain over a long period of time
48 Births and deaths, and emigration and immigration
49 Demography
50 5202
51 A crop of wheat or barley, or a forest of pine trees
52 A weed-killer
53 Because different crops are affected by different pests and take different quantities of minerals from the soil
54 Genetic engineering: genes can be transferred from one species to another; selective breeding: gene transfer is between members of the same species only
55 Malnutrition: suffering caused by a diet which lacks key nutrients; starvation: suffering caused by a diet which lacks all nutrients, particularly those providing energy
56 Because the population is increasing, and industrialisation is increasing
57 Land which is barely able to support crops
58 DDT is an insecticide which is harmful because it is not bio-degradable

59 Examples include introduction of vaccination, use of antibiotics and antiseptics

60 An organism which causes disease

61 An increase in dissolved minerals/fertiliser in the water, e.g. nitrates or phosphates

62 For the manufacture of proteins and nucleic acids

63 By absorbing nitrate salts from the soil

64 By eating plant or animal protein

65 Nitrogen-fixing bacteria

66 Examples include peas, beans or clover

67 They remove the nitrogen and return it to the atmosphere

68 The fertiliser encourages an increase in biomass in the water ecosystem, and this in turn leads to increased death and decay. The large populations of decomposers remove oxygen from the water

69 The control is used to prove that any changes result from the action of the factor under investigation, and not something else

70 Replicates improve the reliability of the results

71 One only

72 Photosynthesis

73 Because of the burning of fossil fuels

74 Because it traps heat energy, rather in the same way as the glass of a greenhouse

75 There is a general increase in average global temperatures

76 Methane

77 Because of the melting of ice caps and glaciers, and because warm water expands

Answers to IQ questions on pages 68-69

Set 1

Cruel and kind are opposites, as are generous and selfish.

A mouse is a mammal. Fear is an emotion.

Set 2

16 (difference in numbers increasing by 1)

5 (add 2, subtract 1)

$(2 \times 4) - 3 = 5$, and $(4 \times 3) - 2 = 10$, so $(2 \times 3) - 4 = 2$

Set 3

t (first letters of numbers)

m (first letters of days of week)

u (every third letter of alphabet)

GLOSSARY

Acetylcholine	A neurotransmitter which stimulates skeletal muscle and is removed by enzyme action
Acoustic coding	A method of placing information in long term memory, simply by saying the word
Acquired immunity	Immunity obtained in a variety of ways, after conception, i.e. not innate immunity
Actin	A thin muscle filament made of protein, found in association with myosin
Activator	A substance essential for enzyme activity
Active immunity	A type of immunity where the body makes its own antibodies
Active site	The part of an enzyme molecule where the catalytic reaction takes place
Active transport	A method of transporting molecules through membranes, against the concentration gradient, using ATP energy
ADH	A hormone produced by the pituitary gland when water content of blood is low
Adrenaline	A hormone produced by the adrenal gland to prepare the body for 'flight or fight'
Algal blooms	Huge populations of algae which result from pollution of waterways by fertiliser or sewage
Allele	An alternative form of a gene, e.g. A, B and O are alleles for the four blood groups
Allergy	An inappropriate response of the immune system to a relatively innocent substance
Alzheimer's disease	A condition of old age characterised by severe memory loss
Amniocentesis	A procedure used to remove amniotic fluid from around the embryo, so that its cells can be examined for genetic defects
Antibiotics	A group of compounds produced naturally by fungi, which kill bacteria
Antibody	A protein produced by B-lymphocytes to defend the body from infection (antigens)
Anticodon	The triplet of bases found on tRNA molecules, which codes for one amino acid
Antigen	A complex compound, usually a protein, which stimulates the antibody response
Artificial immunity	Immunity obtained by vaccination
Artificial insemination	Insertion of sperm into the uterus/fallopian tubes, by injection, not by sexual intercourse
ATP	A high-energy compound used to drive chemical reactions in the body
Atrio-ventricular node	An area of the heart which co-ordinates the contraction of the ventricles

Autoimmunity	An immune response of the body against its own cells
Autonomic nervous system	Part of the nervous system which controls automatic functions such as heart rate
Autosome	A chromosome which is not a sex-chromosome. We have 44 in each body cell
Axon	A nerve fibre which carries impulses away from a neurone
Bile	A compound produced by the liver, containing bilirubin. It contains no enzymes
Bilirubin	A breakdown product of red blood cells, formed by the liver and found in bile
B-lymphocyte	A white blood cell which produces antibodies
Bowman's capsule	A small cup-shaped structure found in kidney nephrons, where filtration takes place
Carrier (genetics)	A person who is heterozygous for a recessive allele which is potentially harmful
Carrying capacity	The population of organisms which an ecosystem can support in the long term
Cell-mediated response	The cell to cell direct response of T-lymphocytes to foreign material
Central nervous system	The brain and the spinal cord
Cerebellum	The part of the brain which co-ordinates muscular movements and balance
Cerebral hemispheres	The two halves of the cerebrum, connected by the corpus callosum
Cerebrum	The main part of the brain, involved in all conscious activities
Chiasmata	Points where paired chromatids intertwine, then break and rejoin, during meiosis
Chromatids	Daughter chromosomes, formed from DNA replication
Chromosomes	Long coiled strands of DNA containing thousands of genes. We have 46 in each body cell
Chunking	A method of grouping information for easier remembering
Cleavage of zygote	The early divisions of a zygote, where cell size is reduced by half at each division
Co-dominance	The situation where two alleles have an equal effect on the phenotype
Codon	A triplet of bases found on messenger RNA; compare with anticodon
Collecting ducts (kidney)	Kidney tubules where ADH acts and which carry urine to the ureter
Colostrum	The breast milk produced the first few days after birth. It is rich in antibodies
Competitive inhibitor	A substance which competes for the active sites of enzyme molecules

Concentration gradient	The difference in concentration of molecules from one place to the next
Contextual cues	Aids to memory which relate in some way to the item which is remembered
Converging pathways	Nervous pathways which merge into one pathway
Coronary artery	The artery which supplies the heart muscle with blood
Corpus callosum	The band of fibres which links the two cerebral hemispheres
Corpus luteum	The 'yellow body' which forms from the follicle after ovulation
Cristae of mitochondria	The inner folded membranes of the mitochondrion where ATP is produced from the cytochrome system reactions
Crossing over (of genes)	The swapping of genes at chiasmata
Cytochrome system	The third stage of respiration which occurs on the cristae
Deamination	The breakdown of amino acids in the liver with the production of urea
Deindividuation	A situation in which a person behaves differently as part of a crowd
Deletion mutation	A mutation in which a base or part of a chromosome is removed
Demography	The study of human population statistics
Denature	To breakdown the normal structure of proteins (usually enzymes), e.g. by excessive heat
Dendrite	A fibre which carries nervous impulses towards a nerve-cell body
Denitrification	The breakdown of nitrogenous compounds by bacteria in the soil to release nitrogen gas
Dependency	The reliance of offspring on their parents
Desertification	The conversion of good quality soil to poor quality, dry soil with little or no organic material
Diastole	The relaxed phase of the heart beat
Diploid	The number of chromosomes found in body cells (46)
Disaccharide	A sugar molecule composed of two monosaccharides, e.g. sucrose
Discontinuous variation	Variation in which there are only a few distinct differences between individuals
Discrimination	The ability to distinguish between different but related stimuli
Diverging nerve pathways	Nervous pathways which branch out from a single pathway
Dizygotic twins	Twins formed from the fertilisation of two different eggs by two different sperm
Double helix	The name given to the spiral shape of the DNA molecule
Down's syndrome	A genetic condition in humans brought about by an extra chromosome number 21

Elaboration	A method of improving LTM by adding detail to the item being remembered
Encoding	The conversion of information into a form which can be stored in memory
Endocytosis	The uptake of substances by infolding of the cell membrane
Endometrium	The lining of the uterus
Endoplasmic reticulum	Layers of membranes found in the cell, often associated with ribosomes (rough ER)
Enzyme	A protein compound which acts as a catalyst in metabolic reactions
Exocytosis	The export of substances involving the fusion of vesicles with the cell membrane
Extinction (behaviour)	Behaviour which disappears as a result of a lack of reward
Fertilisation	The fusion of the male and female gametes – sperm and eggs – to form a zygote
Fetus	The young human, after around the first 7 weeks of life
Fluid mosaic model	The model of the cell membrane to show a phospholipid bi-layer and a mosaic of proteins
Follicle stimulating hormone	A pituitary hormone which stimulates the development of a ball of cells round the ovum
Gall bladder	Associated with the liver; bile is stored in it
Gamete	A sex cell, e.g. a sperm or egg
Gene	A piece of DNA which codes for a protein
Gene locus	The place on a chromosome where a gene is found
Generalisation (behaviour)	Behaviour in which there is the same response to a variety of related but different stimuli
Genetic screening	The checking of family histories or genotypes for genetic abnormalities
Genotype	The genetic makeup of an organism, usually expressed as letters of the alphabet
Gestation	The period of time during which an embryo/fetus develops in the womb (9 months for humans)
Glomerulus	A knot of capillaries found in the Bowman's capsule of a kidney nephron
Glucagon	A pancreatic hormone which stimulates the conversion of glycogen to glucose
Glycerol	A component of a fat molecule
Glycogen	A polysaccharide storage compound found in the liver and muscles, similar to starch

Glycolysis	The first stage of respiration in which glucose is broken down to pyruvic acid
Golgi body	A cell organelle involved in packaging and preparing substances for excretion
Golgi vesicle	A tiny globule of material budded off from a Golgi body
Graafian follicle	A ball of cells which surrounds the developing egg in the ovary
Growth hormone	A pituitary hormone which stimulates growth
Haemoglobin	A red-coloured protein found in red-blood cells, which carries oxygen
Haploid	The name given to gametes which only have one set of chromosomes (23)
Heterozygous	Having two different alleles, e.g. T and t
Homeostasis	The maintenance of the steady state
Homologous chromosomes	Chromosomes which match because they have the same gene loci
Homozygous	Having two alleles the same, e.g. T and T or t and t
Humoral response	The production of antibodies by B-lymphocytes
Hypothalamus	Linked to the pituitary gland, it monitors body temperature and blood concentration amongst many of its functions
Hypothermia	The drop in body temperature below 35°C, leading to unconsciousness and death
Identical twins	Twins formed from the break-up of a single zygote at an early stage in development
Identification	Behaviour in which an individual copies the behaviour of someone they admire
Immunity	The ability to resist infection by pathogens
Implantation	The embedding of a developing embryo in the endometrium
In vitro **fertilisation**	Fertilisation in glass (e.g. test-tube), outside the body of a female
Incomplete dominance	Dominance in which neither allele dominates the other completely and both alleles influence the phenotype when present together
Independent assortment	The random alignment of chromosome pairs along the equator of the cell during the first division of meiosis
Infant attachment	The emotional binding of a child to its parents during the early stages of childhood
Inhibitor (enzyme)	A compound which limits enzyme activity
Innate behaviour	Behaviour which is inborn
Innate immunity	Immunity which is inborn
Insertion mutation	A mutation in which one or more bases are added to a gene sequence

Insulin	A pancreatic hormone which stimulates the conversion of glucose to glycogen
Intensive farming	Farming in which animals are reared in large numbers in one specially designed area to increase productivity
Internalisation	A form of behaviour in which a person believes in something due to logical argument or persuasion
Interstitial cells	Cells found between the seminiferous tubules of the testes, which produce testosterone
Intrinsic factor	A chemical produced by the stomach which is required for the absorption of vitamin B_{12}
Islets of Langerhans	Cells in the pancreas which produce insulin and glucagon
Karyotype	The characteristic number, shape and size of a set of chromosomes
Krebs cycle	The second stage of respiration in which carbon and hydrogen are removed from a variety of compounds
Lacteal	A central tube found in villi, which transports fats
Lactic acid	The end-product of anaerobic respiration, which accumulates in the body during heavy exercise
Limbic system	A part of the brain involved in memory storage
Lipase	An enzyme which digests fats
Lipids	Fatty substances such as fats, oils, steroids and cholesterol
Long term memory (LTM)	Anything stored in the brain for longer than around 30 seconds can be regarded as LTM
Loop of Henlé	Part of the kidney nephron which is involved in water and salt transport
Luteinising hormone (LH)	A pituitary hormone which stimulates ovulation
Lymph nodes	Glands of the lymph system containing many active lymphocytes
Lymph system	A part of the circulatory system involved in the collection of excess tissue fluid
Lymphatic	A lymph capillary
Lymphocyte	A white blood cell involved in combating infection
Lysosome	An organelle containing powerful enzymes capable of digesting bacteria and other cell debris
Macrophage	A white blood cell which can carry out phagocytosis to remove bacteria and other debris
Malnutrition	A condition in which the diet lacks certain nutrients
Marginal land	Land which is barely able to support grazing animals or crops
Matrix of mitochondrion	The central part of the mitochondrion in which the Krebs cycle takes place

Maturation	The stage of development of an individual
Medulla oblongata	Part of the brain (the brain stem) where basic bodily functions are controlled, e.g. heart rate
Meiosis	Division of the nucleus during the formation of gametes, in which genes are swapped and the chromosome number is halved
Memory span	The capacity of the short term memory. Usually around 7 pieces of data
Menstrual cycle	The monthly female reproductive cycle during which ovulation and menstruation take place
Messenger RNA (mRNA)	A molecule which carries the genetic code from the DNA to the ribosomes
Methane	A gas produced as a result of the decaying activity of bacteria in marshes and in the guts of animals
Micrometre/micron	A thousandth of a millimetre (μm)
Mitochondrion	A cell organelle which produces ATP as a result of aerobic respiration
Mitosis	Division of the nucleus of body cells in which the chromosome number is maintained
Mnemonic	A memory aid, e.g. a poem or phrase of associated words or letters
Monoculture	An area containing plants of one species, e.g. a field of wheat
Monosaccharide	A simple sugar, e.g. glucose
Monozygotic twins	Twins formed from the break-up of a single zygote at an early stage in development
Mortality rate	Death rate, often expressed as number of deaths per thousand of the population
Motor areas of the brain	Parts of the cerebrum which control muscle movement
Motor neurone	A nerve cell which carries stimuli from the brain to the muscles
Multiple alleles	A situation in which one of three or more alleles can be found at a particular gene locus
Mutagen	A factor which causes mutations to occur, e.g. a chemical or certain forms of radiation
Mutation	An unplanned, random change to the genetic code, usually harmful
Myelin sheath	A fatty layer of membranes surrounding axons, which speeds up impulse transmission
Myosin	A thick muscle filament made of protein, found in association with actin
NAD	A compound which carries hydrogen in cell metabolism, e.g. during respiration
Natural immunity	Immunity which is acquired by natural means, e.g. catching a cold

Negative feedback	A mechanism for self-regulation in which the increase of a factor sets off a mechanism which brings about its decline, and vice versa
Nephron	A microscopic kidney unit in which plasma is filtered and then useful substances are reabsorbed
Nerve cell body	The part of the nerve cell which contains the main cell organelles, and from which fibres radiate
Neuromuscular junction	A junction between a neuron and a muscle
Neurone	A nerve cell
Neurotransmitter	A compound which transmits a nervous impulse by diffusing across a synapse from one neuron to another
Nitrifying bacteria	Bacteria which convert ammonia (from decay) into nitrites and nitrates
Nitrogen fixing bacteria	Bacteria which trap atmospheric nitrogen and convert it to other nitrogenous compounds
NMDA (brain)	A type of receptor found at synapses
Non-competitive inhibitor	An enzyme inhibitor which attaches itself to part of an enzyme and alters the shape of its active site
Non-disjunction mutation	A mutation in which one or more chromosomes move to the wrong pole of the cell
Non-verbal communication	Communication not involving the spoken word, e.g. smiling
Noradrenaline	A neurotransmitter which is reabsorbed without being degraded by enzymes
Nucleic acid	A general term for DNA or RNA
Nucleolus	Part of the nucleus where RNA is manufactured
Nucleotide	A molecular unit of nucleic acid which consists of a base, phosphate and sugar molecule
Oestrogen	A hormone involved in the oestrous cycle, with a variety of functions
Oestrous cycle	see *Menstrual cycle*
Organisation (memory)	The arrangement of items to be memorised into related groups
Osmoregulation	The maintenance of a constant concentration of dissolved substances in the body fluids
Osmosis	The diffusion of water through a membrane
Oviduct	The tube connecting the ovary to the uterus, in which fertilisation takes place
Ovulation	The release of an egg from the ovary
Oxygen debt	The build up of lactic acid during heavy exercise, which has subsequently to be removed

Oxygen dissociation curve	A graph showing the ability of haemoglobin to carry oxygen under different conditions
Oxytocin	A pituitary hormone which stimulates the contraction of the uterus at birth, and the expression of milk during breast feeding
Pacemaker	An area of the heart, in the right atrium, which controls heart rate
Pancreas	An organ which produces insulin, glucagon and many digestive enzymes
Parasympathetic nervous system	Part of the autonomic nervous system which prepares the body for inactivity and rest
Passive immunity	Immunity in which the body receives ready-made antibodies
Pathogen	An organism which causes disease, e.g. bacterium, virus or fungus
Peptide bonds	Bonds which link amino acids to form the primary structure of protein
Peripheral nervous system	The nerves which radiate from the brain and spinal cord to all parts of the body
Peristalsis	A muscular process which moves food along the gut
Phagocytosis	A cellular process in which tiny particles are engulfed by the folding of the cell membrane
Phenotype	The appearance of an organism, often with respect to its genes
Phenylketonuria (PKU)	A potentially fatal genetic condition in which an amino acid is not metabolised
Pinocytosis	A cellular process in which tiny quantities of liquid are engulfed by the folding of the cell membrane
Pituitary gland	A gland found on the underside of the brain, which produces many hormones
Placenta	A disc of tissue linking the unborn child to its mother via the umbilical cord
Plasma (blood)	The fluid portion of the blood, containing many dissolved substances
Plasma membrane	The membrane which surrounds a cell
Plasticity (of brain)	Its ability to transfer a degree of function from one part to another, after damage
Polygenic inheritance	Inheritance in which a characteristic is affected by many genes
(Poly)peptide	A short string of amino acids
Polysaccharide	A carbohydrate such a starch or glycogen, made up of many simple sugar molecules
Population pyramid	A pyramid to show the distribution of different age groups of a population
Postnatal development	Development after birth
Postnatal screening	Checking for abnormalities in a child after birth

Prenatal development	Development before birth
Primary response	The response of the immune system to its first experience of a particular antigen
Progesterone	A hormone produced by the corpus luteum, involved in the menstrual cycle
Prolactin	A pituitary hormone which stimulates the production of milk by the breasts
Prostate	A gland of the male reproductive system which contributes fluid to the semen
Protein	An organic compound containing the elements C, H, O and N
Proximal tubule	The first part of the kidney nephron, just below the Bowman's capsule
Receptor	A molecule found in a synapse which receives a neurotransmitter, e.g. NMDA
Recessive	An allele which only expresses itself in the phenotype when present with another of the same
Reflex actions	Automatic, quick responses to stimuli
Rehearsal (memory)	Repeating something to be memorised a number of times to aid storage
Reinforcement (behaviour)	The rewarding of behaviour to encourage its repetition
Replication (DNA)	The duplication of a DNA molecule
Retrieval (memory)	The recovery of memorised information from the brain
Rhesus blood group	A blood group which is inherited and which has to be matched for safe transfusion
Ribosome	A cell organelle involved in protein synthesis
RNA (mRNA and tRNA)	Nucleic acids involved in protein synthesis
Secondary response	The response of the immune system to its second experience of a particular antigen
Selective breeding	Breeding in which organisms with desirable characteristics are crossed
Selectively permeable	Of a membrane which allows some substances to pass through and not others
Self antigens	Antigens which belong to the body cells. They are unique to each individual
Semantic coding (memory)	Memory of the general gist or meaning of something, rather than of its detail
Semilunar valve	A valve found in the aorta and pulmonary vein which stops the blood flowing back into the heart
Seminal fluid (semen)	Fluid which carries the sperm from the testes out of the penis during ejaculation

Seminal vesicle	A gland of the male reproductive system which contributes fluid to the semen
Seminiferous tubules	Tubules of the testes in which sperm are manufactured
Sensory areas of brain	Parts of the cerebrum which receive information from sensory organs
Sensory neurone	A nerve cell which brings information to the brain from the sense organs
Serial position effect	The memorisation of a list of items in which items at the beginning and the end are best remembered
Sex chromosomes	The X and Y chromosomes which dictate the sex of an organism. XX = female and XY = male
Sex linked	Of genes which are found on the X or Y chromosomes
Shaping (behaviour)	The rewarding of behaviour which approximates to the ultimate required behaviour
Short term memory (STM)	Memory of around 7 items which enables sense to be made of speech and the written word
Sino-atrial node (SAN)	see *Pacemaker*
Social facilitation	The improvement of performance in the presence of others – as competitors or as audience
Somatic cell	A body cell, as opposed to a gamete
Somatic nervous system	The part of the nervous system involved in the conscious control of activities
Specificity (antibodies)	The ability of only one type of antibody to attach to one type of antigen
Specificity (enzymes)	The ability of only one enzyme to catalyse one reaction
Sperm mother cell	The diploid cells which undergo meiosis to form sperm
Sphygmomanometer	An instrument used to measure blood pressure
Spindle fibres	Fibrous proteins which pull chromosomes/chromatids apart during mitosis/meiosis
Substitution mutation	A mutation in which one base is substituted by another
Substrate	The substance on which an enzyme acts
Sympathetic nervous system	Part of the autonomic nervous system which prepares the body for action
Synapse	A tiny gap between adjacent neurones, across which a neurotransmitter diffuses
Systole	The contracting phase of the heart beat
Testosterone	A sex hormone which promotes male characteristics and sperm production
Thyroxine	A hormone which influences the metabolic rate

Tissue fluid	Fluid found between the cells of the body, derived from blood plasma
T-lymphocyte	A white blood cell involved in protection of the body from infection
Transcription (DNA)	The transfer of the genetic code from the DNA to mRNA
Transfer RNA (tRNA)	RNA which carries amino acids to the ribosomes for assembly to protein
Translation (RNA)	The conversion of the base sequence to a sequence of amino acids in the manufacture of protein
Translocation mutation	A mutation in which a piece of chromosome is transferred from one to another
Tricuspid valve	The atrio-ventricular valve on the right-hand side of the heart
Triplet code	Any three bases which code for an amino acid
TSH (hormone)	The hormone which stimulates the activity of the thyroid gland
Ultrafiltration	Filtration of blood plasma in the kidney nephron
Umbilical cord	The blood vessels which link the baby to the placenta
Urea	A poisonous end-product of deamination, removed from the blood by the kidneys
Vaccination	The introduction of a weakened pathogen by inoculation, to produce an immune response
Vasoconstriction	The constriction of blood vessels to reduce the flow of blood
Vasodilation	The dilation of blood vessels to increase the flow of blood
Vesicle (Golgi)	A tiny globule of material, budded off from a Golgi body
Villi (intestinal)	Tiny projections in the intestine, which increase the surface area for absorption
Visual coding (memory)	Memory of a picture of something, e.g. a person's face
Vitamin B$_{12}$	A vitamin required for the manufacture of red blood cells
Zygote	A fertilised egg

HIGHER

Human Biology
course notes 2nd edition

SECOND EDITION
2

✕ Andrew Morton ✕

This second edition is specifically designed to support students of Higher Human Biology and contains everything you need to know to succeed in your exam:

- **Delivers complete coverage** of the three mandatory course units
- **Completely refreshed text** ensures that the most up-to-date issues are covered
- **Extensive full colour diagrams, illustrations and information boxes** stimulate and support learning and improve recall for exams
- **New Tips for the exam** help students improve question response
- **New topic-specific questions** help students reinforce their understanding
- **Glossary of essential terms** provides students with quick and easy reference material

With Higher Human Biology Course Notes, you can enjoy your studies – and be confident of getting a really good grade in your exam!

Other Human Biology titles available from Leckie & Leckie include:
Questions in Higher Human Biology   978-1-898890-17-1

Scotland's leading educational publishers

Leckie & Leckie
4 Queen Street, Edinburgh EH2 1JE
T: **0131 220 6831** F: **0131 225 9987**
E-mail: **enquiries@leckieandleckie.co.uk**
Web: **www.leckieandleckie.co.uk**

Mixed Sources
Product group from well-managed
forests and other controlled sources
www.fsc.org Cert no. CU-COC-809367
© 1996 Forest Stewardship Council

ISBN 978-1-84372-489-6

9 781843 724896

Leckie & Leckie is a division of
Huveaux Plc.

Migration and social mobility

The life chances of Britain's minority ethnic communities

Lucinda Platt